# Cause and Effect, Conditionals, Explanations

*Essays on Logic as the Art of Reasoning Well*

Richard L. Epstein

Illustrations by Alex Raffi

Advanced Reasoning Forum

COPYRIGHT © 2011 Richard L. Epstein.

ALL RIGHTS RESERVED. No part of this work covered by the copyright hereon may be reproduced or used in any form or by any means—graphic, electronic, or mechanical, including photocopying, recording, taping, Web distribution, information storage and retrieval systems, or in any other manner—without the written permission of the author.

The moral rights of the author have been asserted.

Names, characters, and incidents relating to any of the characters in this text are used fictitiously, and any resemblance to actual persons, living or dead, is entirely coincidental. *Honi soit qui mal y pense.*

For more information contact:
    Advanced Reasoning Forum
    P. O. Box 635
    Socorro, NM 87801 USA
    www.AdvancedReasoningForum.org

ISBN 978-0-9834521-0-2

# Cause and Effect, Conditionals, Explanations

Richard L. Epstein

Preface

**Background: Claims, Inferences, Arguments** . . . . 1

**Reasoning about Cause and Effect** . . . . . . . . 13

**The Directedness of Emotions** . . . . . . . . . . 95

**Conditionals** . . . . . . . . . . . . . . . . . . 101

**Explanations** . . . . . . . . . . . . . . . . . 127

Bibliography . . . . . . . . . . . . . . . . . 177

Index . . . . . . . . . . . . . . . . . . . . 189

# *Essays on Logic as The Art of Reasoning Well*
## Contents of Other Volumes in the Series

**REASONING in SCIENCE and MATHEMATICS**
Models and Theories
Experiments
On Mathematics

**PRESCRIPTIVE REASONING**
Reasoning with Prescriptive Claims
Prescriptive Models
Vagueness, Supervaluations, and the Method of Reflective Equilibrium
Rationality

**The FUNDAMENTALS of ARGUMENT ANALYSIS** (2013)
Arguments
Base Claims
Fallacies
Generalizing
Analogies
Induction and Deduction
Rationality
Truth
Probabilities
On Metaphysics

**REASONING and FORMAL LOGIC** (2013)
Valid Inferences
The Metaphysical Basis of Logic
A General Framework for Semantics for Propositional Logics
Why Are There So Many Logics?
Truth
Vagueness
On Translations
Three Questions about Logic
Language, Thought, and Meaning
Reflections on Some Technical Work in Formal Logic
    Gödel's Theorem
    Categoricity with Minimal Metaphysics
    On the Error in Frege's Proof that Names Denote
The Twenty-first or "Lost" Sophism on Self-Reference of John Buridan

# Preface

This series of books presents the fundamentals of reasoning well, in a style accessible to both students and scholars. The text of each essay presents a story, the main line of development of the ideas, while the footnotes and appendices place the research within a larger scholarly context. The essays overlap, forming a unified analysis of reasoning, yet each essay is designed so that it may be read independently of the others. The topic of this volume is the evaluation of reasoning about cause and effect, reasoning using conditionals, and reasoning that involves explanations.

The first essay summarizes material that can be useful as background from *The Fundamentals of Argument Analysis* in this series.

The essay "Reasoning about Cause and Effect" sets out a way to analyze whether there is cause and effect in terms of whether an inference from a claim describing the purported cause to a claim describing the purported effect satisfies specific conditions. Different notions of cause and effect correspond to placing different conditions on what counts as a good causal inference. An application of that method in "The Directedness of Emotions" leads to a clearer understanding of the issue whether every emotion need be directed at something.

In the essay "Conditionals" various ways of analyzing reasoning with claims of the form "if . . . then . . ." are surveyed. Some of those uses are meant to be judged as inferences that are not necessarily valid, and conditions are given for when we can consider such an inference to be good.

In "Explanations" verbal answers to a question why a claim is true are evaluated in terms of conditions placed on inferences from the explaining claims to the claim being explained. Recognizing that the direction of inference of such an explanation is the reverse of that for an argument with the very same claims is crucial in their evaluation. Explanations in terms of functions and goals are also investigated.

In a companion volume *Reasoning in Science and Mathematics* causal reasoning and explanations are connected to the use of models and theories in science. In particular the nature of causal laws is discussed in the context of theories. The study of laws and confirmation is also analyzed in the essay "Generalizing" in *The Fundamentals of Argument Analysis* in this series.

Reasoning well about cause and effect, understanding how to use conditionals, employing and evaluating explanations—these are skills that can benefit us not only in our daily lives, but in science and our search for fundamentals of our knowledge and experience. Come, let us reason together.

> For never yet has any one attained
> To such perfection, but that time, and place,
> And use, have brought addition to his knowledge;
> Or made correction, or admonished him,
> That he was ignorant of much which he
> Had thought he knew; or led him to reject
> What he had once esteemed of highest price.
> 
> > Attributed to the old man in the comedy by
> > William Harvey, *De generatione animalum*

# Acknowledgments

Many people have helped me over the years I have been working on this material. William S. Robinson and Fred Kroon, in particular, have given much of their time and thought to suggestions that have improved the work. The late Benson Mates was a major inspiration for much of the effort to clarify my ideas.

Charlie Silver, Branden Fitelson, Peter Eggenberger, Carolyn Kernberger, and the members of the Advanced Reasoning Forum helped me a lot in the initial stages of the work, while Michael Rooney offered comments on the final draft. Throughout I have benefited much from the editorial advice of Peter Adams.

Much that is good in this book comes from the generous help of these people, to whom I am most grateful. The mistakes are mine, all mine.

# Publishing history of the essays in this volume

The essay "The Directedness of Emotions" is new and grew out of an e-mail dialogue with Fred Kroon. The other essays are revisions, often quite substantial, of material in the corresponding sections of my *Five Ways of Saying "Therefore"*.

The first presentation of the relation of arguments and explanations was at a talk at the Second Conference on Logic and Reasoning of the Advanced Reasoning Forum that was held in Bucharest, Romania, sponsored by New Europe College; it was published as "Arguments and Explanations" in the *Bulletin of Advanced Reasoning and Knowledge*, vol. 1, 2001, pp. 3–17.

The first exposition of the method of evaluating causal claims as inferences appeared in the first edition of my *Critical Thinking*.

# Background: Claims, Inferences, Arguments

## Claims

*Claims*  A *claim* is some written or uttered declarative sentence that we agree to view as true or false, but not both.

The word "uttered" is meant to include silent uttering to oneself. From now on I'll use "utterance" to include writing, too.

We do not need to make a judgment about whether a sentence is true or whether it is false in order to classify it as a claim. A claim need not be an *assertion*: an sentence put forward as true by someone.

Some say that claims only represent things that are true or false, namely, abstract propositions or thoughts. But it's utterances we use in reasoning together, and we can focus on those, as representatives, if you like, of abstract propositions or thoughts.

The word "agree" in the definition of "claim" suggests that it is a matter of convention whether we take a sentence to be a claim. But almost all our conventions, agreements, and assumptions are implicit. Our agreements may be due to many different reasons or causes, including perhaps that there are abstract propositions.

Often when we reason we identify one utterance with another, as when Dick says "Ralph is not a dog" and later, when Suzy thinks about it, she says "I agree. Ralph is not a dog." We do so when we believe the utterances are equivalent for all our purposes in reasoning.

*Equivalent claims*  Two claims are equivalent for our purposes in reasoning if no matter how the world could be, the one is true if and only if the other is true.

I will often assume such equivalences without explicitly saying so.

Often what people say is *too vague* to take as a claim: there's no single obvious way to understand the words, as when someone says

## 2  Background

"This is a free country." Yet, since everything we say is somewhat vague, it isn't whether a sentence is vague, but whether it's too vague, given the context, for us to agree to view it as true or false. In an auditorium lit by a single candle some parts are clearly lit and some are clearly dark, even if we can't draw a precise line where it stops being light and starts being dark. The *drawing the line fallacy* is to argue that if you can't make the difference precise, there's no difference.

## Inferences

We reason in order to discern whether certain claims are true. But we also reason to discern whether a particular claim follows from one or more other claims. We might not know whether those other claims are true. But were they true, would the truth of this other claim follow? At the basis of all reasoning is the notion of one claim following from one or more other claims.

---

***Inferences***  An *inference* is a collection of claims, one of which is designated the ***conclusion*** and the others the ***premises***, which is intended by the person who sets it out either to show that the conclusion follows from the premises or to investigate whether that is the case.

---

Whether some claims constitute an inference depends on the intent of the person uttering them. Sometimes people indicate that intention by using certain words to indicate that a claim is meant as a premise, or as a conclusion, or to indicate whether he or she believes the claim.

*Conclusion indicators*
>hence; therefore; so; thus; consequently;
>we can then show that; it follows that

*Premise indicators*
>since; because; for; in as much as; given that;
>suppose that; it follows from; on account of; due to

*Indicators of speaker's belief*
>probably; certainly; most likely; I think

These and many other indicator words are not part of a claim but show our intent in using the claim in a particular way.

In order to investigate the idea of a conclusion following from premises we make some definitions.

***Valid, strong, and weak inferences***   An inference is *valid* if it is impossible for the premises to be true and conclusion false at the same time and in the same way.

An inference is *strong* if it is possible but unlikely for the premises to be true and conclusion false at the same time and in the same way. An inference is *weak* if it is neither valid nor strong.

The classification of invalid inferences is on a scale from strongest to weakest as we deem more or less likely the possibilities we consider in which the premises are true and conclusion false.

---

For example, the following is a valid inference:

Maria is a widow.
So Maria was married.

We do not know if the premise is true, but if it is, then the conclusion is not false. In this case the conclusion surely follows from the premise.

The following is valid, too:

All dogs bark.
Spot is a dog.
So Spot barks.

Here we know that the first premise is indeed false: Basenjis can't bark, and some dogs have had their vocal cords cut. But it's not the truth or falsity of the premises and conclusion that determines whether an inference is valid, strong, or weak; rather, it is the ways in which the premises and conclusion could be true or false. In any way that the premises of this inference might be true, the conclusion would be also.

In contrast, the following inference is strong:

Almost all dogs bark.
Ralph is a dog.
So Ralph barks.

If we know no more about Ralph than that he is a dog, then any way in which the premises could be true and conclusion false is unlikely, for we know how rare those are. In this case, too, we say that the conclusion follows from the premises, though there is no certainty, no "must" in that. It is only that relative to what we know, it seems to us very unlikely that the premises could be true and conclusion false.

The following, however, is weak:
Louise is a student.
So Louise isn't married.

There lots of ways the premise could be true and conclusion false: for all we know, Louise might be forty years old with a husband and child.

Our evaluation of the strength of an inference is relative to what we believe. "Likely" means "likely to us." But typically the scale from strong to weak is not so completely relative to a particular person that there is no hope we can agree on the strength of inferences. Suppose we disagree. I find a particular inference strong, and you find it weak. If we wish to reason together, you should describe to me a way the premises could be true and the conclusion false that you think is not unlikely. That may depend on knowledge you have of how the premises could be true which I do not have, but once you've made that explicit we can agree or disagree that there is such a possibility. The only issue, then, would be whether we agree that the possibility is likely. Sometimes we can't come to a clear determination, but further examination will leave us with a clearer understanding of what our differences in evaluation are, based on more than just whim. When the beliefs involved in determining the strength of an inference are made explicit, determining the inference to be strong or weak is far more likely to be a shared judgment.

In sum, we say that the conclusion of an inference *follows from* the premises if the inference is valid or strong.

## Arguments

The paradigmatic use of inferences is in attempts to convince someone that a claim is true.

---

**Arguments** An *argument* is an inference that is intended by the person who sets it out to convince someone, possibly himself or herself, that the conclusion is true.

---

Arguments are attempts to convince, whether someone tries to convince you, or you try to convince someone else, or you try to convince yourself. But that does not mean that the criterion for whether an argument is good is whether the argument actually does convince. If I'm drunk, you may give me an excellent argument that my driving home is dangerous; though I remain unconvinced, the argument is no worse. A politician may make a bad argument that you should vote for her, but though you may be convinced, that does not mean the argument is good. Perhaps other ways to convince, such as entreaties, exhortations, sermons, or advertisements, can be judged by how well they convince, but that is not a criterion for judging attempts to establish the truth of a claim. A *good argument* is one that gives us good reason to believe the conclusion. But what does "good reason" mean?

If an argument is to give us good reason to believe its conclusion, we should have good reason to believe its premises, for from a false claim we can reason as easily to a false conclusion as a true one.

> The Prime Minister of England is a dog. All dogs have fur.
> So the Prime Minister of England has fur. (false conclusion)
>
> The Prime Minister of England is a dog. All dogs have a liver.
> So the Prime Minister of England has a liver. (true conclusion)

It seems, then, that a good argument should have true premises. But consider:

> There are an even number of stars in the sky.
> So the number of stars in the sky can be divided by 2.
>
> There are an odd number of stars in the sky.
> So the number of stars in the sky cannot be divided by 2.

One of these has a true premise, but we cannot tell which. A standard that gives us no way to evaluate arguments is not part of the art of reasoning well. Rather, for an argument to be good we must have good reason to believe its premises. We might, though, have good reason to believe the premises and not be aware of that. For an argument to be good, we need to recognize that we have good reason to believe the premises and actually believe them, for otherwise what convincing is done has no basis in our beliefs.

---

***Plausible claims*** A *claim is plausible* to a particular person at a particular time if:
- The person has good reason to believe it.
- The person recognizes that he or she has good reason to believe it.
- The person believes it.

A claim that is not plausible is *implausible* or *dubious*.

---

The classification of claims as plausible or implausible is on a scale from the most plausible, ones we recognize as true, to the least plausible, those we recognize as false. Though we do not have precise measures of plausibility, we can often compare the plausibility of claims and by being explicit about our background we can usually agree on whether we will take any particular claim to be plausible. If we did not think that we can share our judgments of what is plausible, we would have no motive for trying to reason together. So if I say a claim is *plausible* without specifying a particular person, I mean it's plausible to most of us now, as I'm writing.

Good reason to believe a claim needn't always be established by reasoning, for then we would have no place to start, no plausible claims that would not require further justification, continuing forever. Some claims we take as plausible because of our personal experience, or our trust in authority, or our beliefs about the nature of the world.

But it's not just that the premises of a good argument have to be plausible. They have to be more plausible than the conclusion, for otherwise they would give us no more reason to believe the conclusion than we had without the argument.

---

**Begging the question**  An argument *begs the question* if it has a premise that is not more plausible than its conclusion.

---

Further, for an argument to be good, the conclusion must follow from the premises. For example, consider:

Richard L. Epstein is the author of this essay.
So Richard L. Epstein is bald.

This argument is weak: there are lots of likely ways the premise could be true and conclusion false. Though you know that the premise is true, it gives no reason to believe the conclusion.

Arguments, being inferences, are classified as valid, strong, or weak, and as with inferences it is valid or strong ones we take as establishing that the conclusion follows from the premises. But do strong arguments give good reason to believe the conclusion?

As an example, consider that last week Dick heard there are parakeets for sale at the mall. He knows that his neighbor has a birdcage in her garage, and he wonders whether the cage will be big enough for one of those parakeets. He reasons:

(‡)  Every parakeet I or anyone I know has seen, or read, or heard about is less than 50 cm tall. So the parakeets on sale at the mall are less than 50 cm tall.

This argument is not valid. A new kind of parakeet that is 1 meter tall might have been discovered in the Amazon; or a new supergrow bird food might have been developed that makes parakeets grow very tall; or aliens might have captured some parakeets and shot them with rays to make them very large; or . . . . But any possibility Dick or we can think of for the premise to be true and conclusion false is unlikely—so unlikely that Dick and we have good reason to believe the conclusion. The argument is strong.

Strong arguments, as in this example, give us good reason to believe the conclusion, or at least good enough reason for our daily lives and, as we'll see in the following essays, for science. Moreover, a strong argument is often better than a valid one with the same conclusion. Replacing the premise of (‡) with "All parakeets are less than 50 cm tall" would yield a valid but worse argument, for that claim is less plausible than the premise of (‡). There is often a trade-off between how plausible the premises of an argument are and how strong the argument is: the less plausible the premises, the stronger the argument.

Arguments that lie in the broad center of the scale and the clearly weak ones are certainly not good. We needn't bother to classify them as bad versus very bad, since any bad argument tells us nothing about the conclusion we didn't already know.

We now have three tests an argument must pass for it to be good.

---

*Necessary conditions for an argument to be good*
- The premises are plausible.
- The argument does not beg the question.
- The argument is valid or strong.

---

These conditions are relative to a particular person, though we can have confidence that they establish an intersubjective standard for the evaluation of arguments.

Whether these conditions are also sufficient is a large topic which is examined in the companion volume in this series *The Fundamentals of Argument Analysis*. In what follows, though, I will generallly treat them as both necessary and sufficient.

## Repairing arguments

In our daily lives we often treat many arguments as good that do not seem to satisfy these conditions. For example, consider:

    Lee:    Tom wants to get a dog.
    Maria:  What kind?
    Lee:    A dachshund. And that's really stupid, since he wants one that will chase a frisbee.

Lee has made an argument, if we interpret rightly what he said: Tom wants a dog that will chase a frisbee, so Tom shouldn't get a dachshund. On the face of it that argument is not strong or valid. Still, Maria knows very well, as do we, that a dachshund would be a bad choice for someone who wants a dog to play with a frisbee: Dachshunds are too low to the ground, they can't run fast, they can't jump, and the frisbee is bigger than they are, so they couldn't bring it back. Any dog like that is a bad choice for a frisbee partner. Lee just left out these obvious claims. But why should he bother to say them?

We normally leave out so much that if we look only at what is said, we will be missing too much. We can and should revise many arguments by adding an unstated premise or even an unstated conclusion.

When are we justified in doing so? How do we know whether we've revised an argument well or just added our own assumptions? To repair arguments that are apparently defective, we must have some standards, for otherwise we will end up putting words in other folks' mouths. Such standards depend on what we can assume about the person with whom we are reasoning or whose work we are reading.

---

***The Principle of Rational Discussion***   We assume that the other person with whom we are deliberating or whose reasoning we are evaluating:

- Knows about the subject under discussion.
- Is able and willing to reason well.
- Is not lying.

---

Often someone with whom we wish to reason does not satisfy these conditions. But when we discover that, then it makes no sense to continue to reason with him or her. We should be educating, or consoling, or pointing out errors.

Background 9

The Principle of Rational Discussion justifies adopting the following guide.

---

***The Guide to Repairing Arguments***  Given an (implicit) argument that is apparently defective, we are justified in *adding* one or more premises or a conclusion if and only if all the following hold:
- The argument becomes valid or strong.
- The premise is plausible and would seem plausible to the other person.
- The premise is more plausible than the conclusion.

If the argument is valid or strong, yet one of the original premises is implausible, we may *delete* that premise if the argument becomes no worse. In that case we say the premise is ***irrelevant***.

---

Given only this Guide, we might try to repair every argument into a good one. That would be wrong, for there are standards for when an argument is unrepairable.

---

***Unrepairable Arguments***  We cannot repair a (purported) argument if any of the following hold:
- There is no argument there.
- The argument is so lacking in coherence that there's nothing obvious to add.
- A premise is implausible or several premises together are contradictory and cannot be deleted.
- The obvious premise to add would make the argument weak.
- Any obvious premise to make the argument strong or valid is implausible.
- The conclusion is clearly false.

---

It's not that when we encounter one of these conditions we can be sure the speaker had no good argument in mind. Rather, we are not justified in making that argument for him or her, for it would be putting words in his or her mouth.

In addition to these conditions for an argument to be unrepairable, a list of other kinds of arguments, called fallacies, have been deemed to

be typically so bad that they, too, are rejected as unrepairable when we encounter them.
   Consideration of two particular kinds of arguments is important for the essays that follow: generalizations and analogies.

## Generalizations

***Generalizations*** A *generalization* is an argument in which we conclude a claim about a group, called the ***population***, from a claim about some part of it, the ***sample***. Sometimes we call the the conclusion the *generalization*. Plausible premises about the sample are called the ***inductive evidence*** for the generalization.

The following are generalizations:

(‡) Every dog I've seen barks.
   So all dogs bark.
   *sample*: The dogs the speaker has seen.
   *population*: All dogs.

Every dog I ever met except one can bark.
So almost all dogs bark.
   *sample*: The dogs the speaker has met.
   *population*: All dogs.

Of dog owners who were surveyed, 98.2% said their dogs bark.
So about 98% of all dogs bark.
   *sample*: The dogs of the pet owners surveyed.
   *population*: All dogs.

The last is called a *statistical generalization* because its conclusion is a statistical claim about the population.
   If we have no reason to think that the sample is similar to the population, then the generalization is worthless, a bad argument. What we want for a good generalization is for the sample to be representative.

***Representative sample*** A sample is *representative* if no one subgroup of the whole population is represented more than its proportion in the population. A sample is ***biased*** if it is not representative.

   The first and second examples at (‡) are bad because we have no reason to think that the dogs the speaker has seen are representative of all dogs. We don't know enough about the sample in the third generalization to make a judgment about whether it's representative.

Random sampling is an important method for getting a representative sample.

---

*Random sampling*  A sample is *chosen randomly* if at every choice there is an equal chance that any of the remaining members of the population will be picked.

---

Random sampling does not guarantee that the sample will be representative. Choosing two students randomly from the 716 at McEpstein High School to interview about their views on gay marriage is not going to give a representative sample. The sample has to be large enough for us to have good reason to think it is representative.

But even if we have confidence that the sample is representative, if it's not studied well then it's no use for concluding anything about the population. Maria asked all but three of the thirty-six people in her class whether they've ever used cocaine, and only two said yes. So she concluded that almost no one in the class has used cocaine. But there's no reason to think that people will answer truthfully to such a question, so her generalization is not good.

---

*Necessary conditions for a generalization to be good*
- The sample is representative.
- The sample is big enough.
- The sample is studied well.

---

These conditions do not establish a different standard from the necessary conditions for an argument to be good. They only spell out in more detail what is required for the argument to be strong.

## Analogies

---

*Analogies*  A comparison becomes *reasoning by analogy* when it is part of an argument: On one side of the comparison we draw a conclusion, so on the other side we say we should conclude the same.

---

For example, consider:

> We should legalize marijuana. After all, marijuana is just like alcohol and tobacco, and those are legal.

The comparison here is between marijuana on the one hand and alcohol and tobacco on the other. The latter are legal. So we should make marijuana legal, too. But no connection between the premise and the conclusion has been supplied besides saying that marijuana "is just like" alcohol and tobacco.

The difficulty in reasoning by analogy is to make clear what we mean by "is just like" or "is the same as" in order to justify the inference in terms of the comparison. Such a justification calls for some general claim under which the two sides of the comparison fall. Often analogies are sketchy, with only the comparison offered, so that their main value is to stimulate us to search for such a general claim. It must be one that relies on the similarities and for which the differences between the two sides of the comparison don't matter. Though that procedure is somewhat more involved than in analyzing many other arguments, it is does not require further necessary conditions for an argument to be good.

## Reasoning backwards

One particular mistake in reasoning is important to note for some of the discussions that follow. For example, Suzy said to Lee:

> All CEOs of computer companies are rich. Bill Gates is a CEO of a computer company. So Bill Gates is rich.

Lee sees that Suzy's argument is valid, and he knows that Bill Gates is rich and that he's a CEO of a computer company. So he reckons that the other premise, "All CEOs of computer companies are rich" is true, too. But he's wrong: there are lots of CEOs of small, struggling computer companies who are not rich. Lee is arguing backwards.

---

*Arguing backwards*  *Arguing backwards* is the mistake of concluding that the premises of an inference are true because the inference is valid or strong and its conclusion is plausible.

---

This concludes the very brief summary of the basics of inference and argument analysis needed for the succeeding essays.

# Reasoning about Cause and Effect

The mystery of cause and effect can be circumvented if not eliminated in our reasoning by using claims to describe purported causes and purported effects and understanding a causal claim as true if and only if the relation between those claims satisfies the conditions for a good causal inference. Different notions of cause and effect correspond to placing different conditions on what counts as a good causal inference. This provides a method of reasoning about cause and effect that is clear and useful in both our ordinary lives and science.

| | |
|---|---|
| Describing causes and effects . . . . . . . . . . | 14 |
| Causation and inference . . . . . . . . . . . . | 17 |
| A minimal notion of cause and effect . . . . . . . | 18 |
| Further conditions on causal inferences . . . . . . | 21 |
| The usual notion of cause and effect . . . . . . . | 27 |
| Cartoon examples . . . . . . . . . . . . . . . | 30 |
| Generalizations in causal reasoning . . . . . . . . | 38 |
| Can a strong causal inference be good? . . . . . . | 42 |
| The normal conditions . . . . . . . . . . . . . | 45 |
|     Summary of issues discussed in the examples . . | 57 |
| Can cause and effect be simultaneous? . . . . . . | 57 |
| Claims about the nature of cause and effect . . . . . | 62 |
| Finding a cause . . . . . . . . . . . . . . . . | 67 |
| Cause in populations . . . . . . . . . . . . . | 70 |
| Causal laws . . . . . . . . . . . . . . . . . | 75 |
| Conclusion . . . . . . . . . . . . . . . . . | 80 |
| Appendix 1  Causes and reasons . . . . . . . . . | 81 |
| Appendix 2  Events . . . . . . . . . . . . . . | 85 |
| Appendix 3  Objective chance . . . . . . . . . . | 89 |
| Appendix 4  Is it new to analyze the relation of cause to effect as an inference? . . . . . . . . . | 90 |

*14   Cause and Effect*

## Describing causes and effects
What is a cause?

> Last night Dick said:
>
> Spot made me wake up.

Spot caused Dick to wake up. But it's not just that Spot existed. It's what he was doing that caused Dick to wake up:

Spot's barking caused Dick to wake up.

So Spot's barking is the cause? What kind of thing is that? The easiest way to describe the cause is to say:

> Spot barked.

The easiest way to describe the effect is to say:

> Dick woke up.

*Causes and effects can be described with claims.*

---

**Causal claims**   A *causal claim* is a claim of the form *X causes* (*caused*) *Y* or a claim that is equivalent to one of that form.

A ***particular*** causal claim is one in which a single claim can describe the (purported) cause, and a single claim can describe the (purported) effect. A ***general*** causal claim is a causal claim that generalizes many particular causal claims.

---

*Example 1*   Spot caused Dick to wake up.

*Analysis*   This is a particular causal claim: the purported cause can be described by the single claim "Spot was barking" and the purported effect by "Dick woke up." We could generalize from this particular cause and effect to, for example:

Very loud barking by a dog near someone when he is sleeping *causes* him to wake up, if he's not deaf.

This is a general causal claim. For it to be true, lots of particular causal claims have to be true.

*Example 2* The police car's siren got Dick to pull over.
    *Analysis* This is a particular causal claim. The purported cause can be described by "The police car had its siren going," and the purported effect by "Dick pulled over."

*Example 3* The speeding ticket Dick got made his auto insurance rate go up.
    *Analysis* This is a particular causal claim. The purported cause is "Dick got a speeding ticket," and the purported effect is "Dick's auto insurance went up."

*Example 4* Speeding tickets make people's auto insurance rates go up.
    *Analysis* This is a general causal claim. For it to be true all particular causal claims like the one in the previous example have to be true.

*Example 5* Penicillin prevents serious infection.
    *Analysis* What is the cause? The existence of penicillin? No, it's that penicillin is administered to people in certain amounts at certain stages of their infections. What is a "serious infection"? This is too vague to count as a causal claim.

*Example 6* Suzy: My boyfriend causes all my problems.
    *Analysis* What is the cause? It's not simply that Suzy's boyfriend exists. It must be what he does. And we have no idea what that is. What is the effect? "Suzy has problems"? Suzy intended to make a causal claim, but what she said is too vague for us to accept as one.

*Example 7* Lack of rain caused the crops to fail.
    *Analysis* Here the purported cause is "There was no rain." The purported effect is "The crops failed." Referring to the period a few years back in the midwest of the United States, this was true.
    Causes do not have to be something active involving change.[1]

---

[1] Perhaps we could say that drought is a change from the normal conditions in the Midwest, but that stretches the notion of change and suggests that we don't have a very clear idea of what counts as change and what as static.

A persistent condition or indeed any description of the world could do. H. L. A. Hart and Tony Honoré say:

> On the one hand it is perfectly common and intelligible in ordinary life to speak of static conditions or negative events as causes: there is no convenient substitute for statements that the lack of rain was the cause of the failure of the corn crop, the icy condition of the road was the cause of the accident, the failure of the signalman to pull the lever was the cause of the train smash. On the other hand the theorist, when he attempts to analyse the notion of a cause, is haunted by the sense that since these ways of speaking diverge from the paradigm where a cause is an event or force, they must somehow be improper. The corrective is to see that in spite of differences between these cases and the simple paradigms, the very real analogies are enough to justify the extension of the causal language to them.[2]

But more simply, we can understand events or static conditions or states of affairs via claims used as descriptions of the world. Descriptions differ in many important respects that matter for cause and effect; for instance, they can be subjective or objective. But little seems to come from viewing them as negative or positive, for many a negative claim can be transformed into a positive one, replacing, for example, "not transparent" by "opaque." We are justified in saying that this example is causal not by analogy, but because, as we will see below, it can be analyzed uniformly with causal claims where the purported cause is "active."

*Example 8*  My fear made me run away from the bear.
  *Analysis*  This is a causal claim, equivalent to "My fear caused me to run away from the bear." It looks like the purported cause is a thing: fear. But we can describe the cause more simply with "I was afraid when I perceived the bear" and the purported effect with "I ran away from the bear."

*Example 9*  The dust made me sneeze.
  *Analysis*  Here it is the dust that is said to be the cause. But it is not just its existence; it is that it existed in this place and this time. The cause can be described with "There was dust in the air here" and the effect with "I sneezed."

*Example 10*  I cause myself to think.

---

[2] *Causation in the Law*, p. 31.

*Analysis* Let's assume the speaker intended this as a particular causal claim, "I cause myself to think right now."

We can describe the effect with "I think." But the cause seems to be a thing, the person who is speaking, not what could be described by a claim. Because there's no obvious way to describe the purported cause with a claim, this example seems very mysterious. Unless we want the example to fall into triviality (the cause, too, being described with "I think"), we should have to take the cause to be something like "I have an act of will." But we should ask the speaker what he or she intended.

*Example 11* Dick's good grade in critical thinking was due to his studying hard.

*Analysis* Here, the purported cause is "Dick studied hard" and the purported effect is "Dick got a good grade in critical thinking." Often it is easier to show that a claim is causal by stating the purported cause and effect than trying to rewrite it as a claim that uses "causes." Whether we accept the example as a causal claim depends on whether we accept that the sentences describing the purported cause or effect are not too vague to be claims.

*Example 12* Because you were late, we missed the movie.

*Analysis* This is a causal claim: we can rewrite it as "Your being late caused us to miss the movie." Causal claims and "becausal" claims can often be interchanged: "A because of B" becomes "B causes A." When we are justified in drawing such equivalences will depend on how we understand the conditions for causal claims to be true.

## Causation and inference

When causes and effects are described with claims, the relation of cause to effect can be analyzed as a relation between claims.

This is not to deny that there is something "out there in the world" that is cause and effect. Perhaps cause and effect is a relation between things in the world, though it seems that it isn't a thing that is a cause, but what it does or does not do, or how it changes or is changed, or just where it is. Perhaps there are events, separate and distinct from things in the world and from the ways we talk about the world, and the causal relation is between them. Or perhaps the world is process without things, and our claims describe parts of that process which are related causally.[3] No matter. By using claims to describe causes and effects we can make that relation less mysterious by analyzing it in terms of a relation between claims.

What is the relation between claims that justifies saying that they describe cause and effect? At the very least, for there to be cause and effect, if the cause is true (happens, occurs), the effect must follow (be true, happen, occur). That is, the relation of cause to effect can be understood as a kind of inference.

## A minimal notion of cause and effect

*The cause and effect are both true*
We won't accept that Spot's barking caused Dick to wake up if Spot didn't bark or Dick didn't wake up. If A describes the cause and B the effect, then both A and B must be true; colloquially, the cause and effect both happened.

*The inference is valid or strong relative to the normal conditions*
It must be impossible or at least very unlikely for the cause to happen and the effect not to follow. That is, the inference must be valid or strong. We would not accept that my washing my car yesterday made it rain today if we can describe ways in which I could have washed my car and it didn't rain today.

But a lot has to be true for it to be impossible for "Spot barked" to be true and "Dick woke up" to be false:

Dick was sleeping soundly up to the time that Spot barked.
Spot barked at 3 a.m.
Dick doesn't normally wake up at 3 a.m.
Spot was close to where Dick was sleeping.
There was no other loud noise at that time.

We could go on forever. But as with arguments, we state what we think is important and leave out the obvious. If someone challenged us, we could add "There was no earthquake at the time"—but we just assume that as part of the normal conditions.

---

**The normal conditions**  The *normal conditions* for a causal claim are the obvious and plausible claims that are needed to establish that the

---

[3] See my *The Internal Structure of Predicates and Names with an Analysis of Reaoning about Process*.

relation between the purported cause and the purported effect is valid or strong.

We can take claims as normal conditions only if they are plausible and make the inference valid or strong.

*The cause makes a difference*
Dr. E has a desperate fear of elephants. So he buys a special wind chime and puts it outside his door to keep the elephants away. He lives in Cedar City, Utah, in North America, at 6,000 feet (1800m) above sea level in a desert, and he confidently asserts "The wind chime causes the elephants to stay away." After all, ever since he put up the wind chime he hasn't seen any elephants. There's a perfect correlation: "Wind chime up, no elephants," Dr. E notes again in his diary on Tuesday.

Why are we sure the wind chime being up did not cause elephants to stay away? Because even if there had been no wind chime, the elephants would have stayed away. Which elephants? All elephants. The wind chime works, but so would anything else. The wind chime doesn't make a difference.

For there to be cause and effect, if the cause hadn't occurred, the effect would not have happened. If A describes the purported cause and B the purported effect, then if A had not been true, B would not be true. If Spot had not barked, Dick would not have woken up. It must be (nearly) impossible for "Spot did not bark" to be true and "Dick woke up" to be false. The inference "B therefore A" must be valid or strong, relative to some assumed claims, perhaps in addition to the normal conditions.

We now have minimal conditions for a notion of cause and effect.

***A minimal notion of cause and effect*** For a particular causal claim to be true, describing the purported cause with a claim A and purported effect with a claim B, the following must hold:

- Both A and B are true.

- Given the normal conditions, the inference from A to B is valid or strong. The claims offered as normal conditions must be plausible and make the inference valid or strong.

- Given the normal conditions and perhaps other plausible claims, the inference from B to A is valid or strong.

A *causal inference* is an inference "A therefore B" that is meant to correspond to, or be identified with, or simply be what is understood by a particular causal claim. A *general causal claim* is true if every causal claim that could be an instance of it is true.

Consider now:

$1 + 1 = 2$ causes $4 + 4 = 8$.

This satisfies the minimal conditions. The minimal notion of cause and effect is so minimal that causes and reasons seem the same. Some have talked of causes as reasons, as you can see in Appendix 1. But even so, if the example is true so is "$4 + 4 = 8$ causes $1 + 1 = 2$." Nothing so far distinguishes cause from effect.

We want to develop conditions on causal inferences so that we'll be justified in saying that a *causal inference is good* just in case the causal claim it is meant to analyze is true. The problem, though, is whether we have any independent standard for a causal claim being true.

Various standards have been offered depending on differing metaphysics, as we'll see below. Different standards have been proposed even relative to the same metaphysics. Other standards, as we'll see, invoke our use of the terms "cause and effect" in ordinary and scientific discourse as justification for when a causal claim should be classified as true. We can compare these various standards by seeing how different conditions arise from or presuppose distinct metaphysics or specific assumptions about the use of language.

We can leave open whether there is an independent standard for cause and effect that correctly reflects the nature of cause and effect in the world, or whether all there is to cause and effect is our propensity to classify parts of our experience in that way. If the former, then the examination of conditions on causal inferences is an examination of the real nature of cause and effect. If the latter, our examination is a way to lead us to agree on how we will classify our experience and use the words "cause and effect" to communicate among ourselves. Whatever conditions we impose must seem natural in either case, either because of our beliefs about the world or because of our use of language.

Arguments differ from causal inferences in one important respect. For an argument to be good it must give good reason to believe its premises, either for a specific person or more generally. That is why the premises of a good argument are required to be plausible, not true. The conditions required for a causal inference are supposed to be ones that correspond to the claim being true, not to our knowing that the claim is true. Even the requirement that the inference must be strong is interpreted by some to be in terms of some objective notion of probability rather than our subjective evaluation relative to our knowledge and beliefs.

Our knowledge and beliefs are important, however, in our classifying causal inferences as good or bad. We evaluate whether claims describing the purported cause and purported effect are true, and that is in terms of whether they are plausible. We evaluate whether the inference is strong in terms of what seems likely to us.

Let's now consider what further conditions might be imposed for a causal inference to be good.

## Further conditions on causal inferences
*Time and space*
The common understanding of cause and effect is that the cause must precede the effect. But then there's something odd in describing a cause with a claim. We say an event (cause) occurred: it happened at some time.

Claims that describe the world are true of particular moments. Thus, "Spot was barking" is true of a particular time: when the world was such as the claim says. But what time was that? An instant? A stretch of time? The claim is vague about that, though we can make it as precise as we wish if needed in our reasoning. "Dick woke up" is true of a later time, though exactly when—at an instant later than the instant of which "Spot was barking" is true, or within the interval of time of which "Spot was barking" is true—is not made precise in the sentence. Still the sentences are clear enough to reason with as claims. So in what follows when I speak of a cause or an effect *happening*, understand that to mean the claim that describes the cause or effect is true of a particular time.

So we add to the conditions on cause and effect that the cause must precede the effect. That is, if A describes the cause and B the effect, A must be true of an earlier time than B. It is the arrow of time that distinguishes cause from effect.[4] If the world were to run backwards, as a movie can run backwards, cause would become effect and effect would become cause, though with respect to different normal conditions.[5] We could not distinguish cause from effect this way if they could be simultaneous, which we'll consider below.

The assumption that the claims describing cause and effect are true of particular times is a very substantial metaphysical constraint. It rules out claims that are about only abstract objects as describing either a cause or effect. Claims such as "1 + 1 = 2" are not true of any specific time. Nor are such claims true of all times. They are (supposing you accept that there are abstract objects) timelessly true. Nor can timeless claims be indexed with times: "1 + 1 = 2 yesterday" is either nonsense or false.

---

[4] If I understand Kant correctly in his *Critique of Pure Reason* at the second analogy of experience in the "Transcendental Analytic" and sections 27–29 of the *Prolegomena*, Kant reverses this step: It isn't time that distinguishes cause from effect, but our prior intuition of cause and effect that gives our perceptions a temporal order. See Robin Le Poidevin and M. MacBeath, *The Philosophy of Time*, pp. 6–8 for a discussion of modern views that temporal relations are definable in terms of causal relations.

More recently, it has been proposed the direction of cause to effect, and hence of time, is that of proceeding from lesser to greater entropy; see for example, Lawrence Sklar, "Up and down, left and right, past and future." But that condition won't help us in reasoning about cause and effect, for it is entirely statistical: over a wide enough region, or perhaps the universe as a whole, there is one direction of order and disorder that can or does determine the direction of time. The causal claim "The cold weather caused the water to freeze" is true, yet frozen water has a lower level of disorder than liquid water. Perhaps there is a theoretical analysis of cause and effect in terms of entropy or some other non-temporal feature of the world. But if so, we must ask how it relates to our need to reason about cause and effect.

[5] See Michael Dummett, "Bringing about the past."

Yet a big problem with abstract objects is if they exist, how can they cause anything? Adopting time as a metaphysical constraint on cause and effect denies that problem; we can only discuss it if we allow that there may be different metaphysical constraints on cause and effect. When someone says that abstract objects are a problem in science because, for example, numbers are causally inert, we hope that he's saying more than that he's defined cause and effect not to involve abstract objects.[6]

Other timelessly true claims which are not about abstract objects are also ruled out as describing causes. For example, "Gold is heavier than water" is not true of any time, but timelessly true. So "Gold is heavier than water" cannot describe the cause of Suzy's ring sinking to the bottom of the lake, though it may be a reason why it sank.

Nor can conditionals or disjunctions serve as causes. "If Spot barked, then Dick woke up; therefore Spot ran away" is not a good causal inference, because we cannot specify a time of which "If Spot barked, then Dick woke up" is true.

What about quantified claims? Consider:

No dog has ever hibernated,
therefore Spot will not hibernate this winter.

If this is a good causal inference, then again causes and reasons are conflated. "No dog has ever hibernated" should be a normal condition, not the cause. Still, it is true of a specific time, namely all of history up to now. To rule it out we could require that no complex claim be a cause: a claim describing a cause must be an atomic claim, or a negation of an atomic claim, or a conjunction of those. To use the notion of atomic claim, though, requires us to specify a particular metaphysics. Typically that is done by taking claims to be formalizable in first-order predicate logic. But that's too narrow, for it allows only things to be involved in causes, not processes or substances.[7] Many say that

---

[6] Compare Bertrand Russell, "On the notion of cause, with applications to the free-will problem," p. 216:

> A causal law, we said, allows us to infer one *thing* (or *event*) from the existence of one or more others. The word "thing" here is to be understood as only applying to particulars, i.e. as excluding such objects as numbers or classes or abstract properties and relations, and including sense-data, with whatever is logically of the same type as sense-data.

[7] See my *The Internal Structure of Predicates and Names*. See also

claims describing cause and effect must be true of specific places, invoking space as a metaphysical constraint on cause and effect. Talk of events taking place at a specific location is supposed to do that (see Appendix 2). But that won't rule out this example, which is true of the earth. *Claims that describe causes and effects are those that—according to whatever metaphysics we adopt—describe the world.* The analysis of cause and effect in terms of inferences will allow for any metaphysics so long as claims that are descriptions of the world are what serve in causal inferences.

*The cause is close in time and space to the effect*
Spot caused Dick to wake up. But Dick and Zoe's neighbor tells them that's not right. It was because of a raccoon in her yard which shares the same fence that Spot started barking. So really, a raccoon entering her yard caused Dick to wake up.

But it was no accident that the raccoon came into their neighbor's yard. She'd left her trash can uncovered. So really the neighbor's not covering her trash caused Dick to wake up.

---

Nicholas Rescher, *Process Metaphysics*. Wesley C. Salmon in "Causality: production and propagation," p. 155, says,

> One of the fundamental changes which I propose in approaching causality is to take processes rather than events as basic entities.

But he takes processes as an addition to the ontology of events (pp. 155–156) and does not give a reduction of events to processes. To base causal analyses on processes he says:

> We need to make a distinction between what I call *causal processes* and *pseudo-processes:* . . . causal processes are those which are capable of transmitting signals; pseudo-processes are incapable of doing so. (Italics and ellipses in original, p. 156.)

But "signals" are effects; so causal processes are those which are causal.

*Cause and Effect* 25

But really it was because Spot knocked over her trash can and the top wouldn't fit, so their neighbor didn't bother to cover her trash. So it was Spot's knocking over the trash can that caused Dick to wake up.

But really, . . . . This is silly. We could go backwards forever. We stop at the first step: Spot's barking caused Dick to wake up. We stop because *as we trace the cause back further it becomes too hard to fill in the normal conditions*.

It's sometimes said that the cause must be close in space and time to the effect. But the astronomer is right when she says that a star shining caused the image on the photograph, even though that star is very far away and the light took millions of years to arrive. The problem isn't how distant in time and space the cause is from the effect. The problem is how much has come between the cause and effect—whether we can specify the normal conditions. When we trace a cause too far back, the problem is that the normal conditions begin to multiply. There are too many conditions for us to imagine what would be needed to establish that it's impossible for the cause to have been true and the effect false.

---

***The cause is close in space and time to the effect***   A purported cause A being *close in space and time* to a purported effect B is best understood as meaning that we can fill in the normal conditions to make the inference from A to B valid or strong.

---

## 26   Cause and Effect

*There is no common cause*

Night causes day.

This is false. If it's false, then since it's a general causal claim, one of its instances is false. Let's consider one I'm making now at noon:

Cause: It was night eight hours ago.
Effect: It is day now.

All the conditions we required for there to be cause and effect are met. But there is a common cause of both "It was night eight hours ago" and "It is day now," namely, "The earth rotates relative to the sun," given some normal conditions.

---

**Common cause**   Given a purported cause A and purported effect B, C is a *common cause* of both A and B means:

- C is a cause of A.
- C is a cause of B.
- A is not needed to make the inference from C to B valid or strong, though A may follow from the normal conditions for that inference. (A does not *intervene* between C and B.)
- B is not needed to make the inference from C to A valid or strong, though B may follow from the normal conditions for that inference. (B does not *intervene* between C and A.)
- C is needed (as a normal condition) to make the inference from A to B valid or strong, if indeed the inference is valid or strong.

---

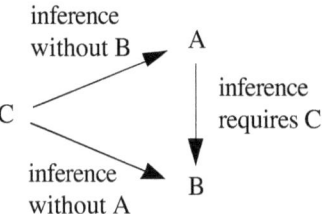

We have to understand the first two conditions here to mean that C satisfies all the conditions to be a cause of both A and B with the possible exception that there is no common cause of C, A, and B, for otherwise we would fall into an infinite regress trying to determine cause and effect.

The inference from "The earth rotates relative to the sun" to "It is day now" does not require as premise "It was night eight hours ago" to be valid or strong, even though that claim is true and follows from the normal conditions. Similarly, for the inference from "The earth rotates relative to the sun" to "It was night eight hours ago" to be valid or strong does not require as premise "It is day now," even though that, too, is true and follows from the normal conditions. Yet the inference from "It was night eight hours ago" to "It is day now" does require "The earth rotates relative to the sun" as a normal condition in order to be valid or strong. Hence, if there must be no common cause for there to be cause and effect, then "Night causes day" is false.

## The usual notion of cause and effect
With these conditions we have the usual notion of cause and effect.

---

*Necessary conditions for cause and effect*  For a particular causal claim to be true, describing the purported cause with a claim A and purported effect with a claim B, the following must hold:

1. Both A and B are true.
2. Given the normal conditions, the inference from A to B is clearly valid or strong.
3. Given the normal conditions and perhaps other plausible claims, the inference from B to A is clearly valid or strong.
4. A is true of an earlier time than B, and both are true of particular places.
5. There is no common cause of both A and B.

Claims offered as normal conditions must be plausible and make the inference valid or strong.

A *causal inference* is an inference "A therefore B" that is meant to correspond to, or be identified with, or simply be what is understood by a particular causal claim.

A general causal claim is true if every causal claim that could be an instance of it is true.

---

Colloquially, conditions (1–5) are:

- A and B both happened.
- It's (nearly) impossible for A to have happened and B not to happen.
- If A hadn't happened, B wouldn't have happened (the cause makes a difference).
- A happened before B happened.
- There is no common cause.

The rubric "The cause is close in space and time to the effect" is just the "clearly" in (2) and (3) as well as a reminder that we have adopted a metaphysics of space and time in our understanding of cause and effect.

It is sometimes said that there is a problem that cause and effect is not "in the world" or that it is in the world but our analyses are faulty because they don't show that.[8] By understanding the causal relation as a kind of inference, cause and effect are as much in the world as any

---

[8] Some even wish to dismiss all talk of cause and effect from the law because questions whether if A had not happened then B would not have happened are not "matters of fact." See, for example, E. Wayne Thode, "The indefensible use of the hypothetical case to determine cause in fact".
H. L. A. Hart and Tony Honoré, *Causation in the Law*, p. 101, say:

> Some later writers, impressed, as logicians always have been, by the fact that singular hypothetical statements of the form "If X which in fact had not happened, Y would not have happened" (contrary-to-the-fact conditionals) are not verifiable or falsifiable in any straightforward way, have claimed that even the determination whether the harm would have happened without the defendant's act cannot be regarded as a factual or policy-neutral question. . . . How can a question about facts which concededly never existed be a question of fact? Examination of decided cases has convinced some critics that the prime determinants of such questions are not factual estimates of probabilities but findings that defendant's action was wrongful, and the kind of harm which came about was of the very kind which the rule it violated was designed to prevent. The further claim that not only is this all that courts in fact often do consider in determining the factual-looking question "Would this harm have occurred but for the defendant's wrongful act?", but all that it should consider represents an uncompromising form of the programme for eliminating causation from the law as an element of responsibility.

of our reasoning. Whether there is something "out there" beyond our reasoning which is the "real" cause and effect in the world is a debate that can continue. In the meantime, we can get on with reasoning about cause and effect in our lives.[9]

I wish we could resolve here what we should take as normal conditions and establish sufficient conditions for a causal inference to be

---

[9] This is not the same as Hume's view that the notion of causal necessity lies entirely in the mind's habit of observing constant conjunction, as will be clearer in the examples below. The necessity (or strength) of the causal relation is as "real" as that of any inference. Hume's rejoinder to his critics in *A Treatise of Human Nature*, pp. 167–168, seems more apt:

> What! the efficacy of causes lie in the determination of the mind! As if causes did not operate entirely independent of the mind, and wou'd not continue their operation, even tho' there was no mind existent to contemplate them, or reason concerning them. Thought may well depend on causes for its operation, but not causes on thought. This is to reverse the order of nature and make that secondary, which is really primary. . . .
>
> I can only reply to all these arguments, that the case is here much the same, as if a blind man shou'd pretend to find a great many absurdities in the supposition, that the colour of scarlet is not the same with the sound of a trumpet, nor light the same with solidity. If we have really no idea of a power or efficacy in any object, or of any real connexion betwixt causes and effects, 'twill be to little purpose to prove, that an efficacy is necessary in all operations.

Ernest Nagel in "In defense of logic without metaphysics," pp. 95–96, agrees that taking causality as "an objective ontological category" serves no purpose in establishing any particular generalization. On the other hand, Jaegwon Kim in "Explanatory realism, causal realism, and explanatory exclusion" gives a defense of "causal realism," the idea that the relation of cause to effect is an objective relation between events in the world. Michael Tooley in *Causation: A Realist Approach* is also a realist about causes:

> First, laws are to be identified with certain contingent, irreducible, theoretically specified relations among universals, and second, that causal relations between states of affairs are also theoretically specified relations. p. 5

He defines "P→Q" as "properties P and Q stand in the relation of direct causal necessitation." It doesn't seem that we can add further metaphysical constraints to the minimal notion proposed here to get his analysis. We might identify a property in his sense pertaining to an individual with an atomic proposition being true; but a negation of an atomic claim could not correspond to a cause, for he says that blue and red are properties, but not-green isn't (p. 6).

30  *Cause and Effect*

good. But at best we can look at many examples and try to refine these conditions. Understanding the causal relation as a special kind of inference does not resolve all problems in our understanding of cause and effect. But it does give us a framework in which to discuss more clearly various points of view about the nature of cause and effect.

Let's now turn to examples that will illustrate and extend our analysis of how to reason about cause and effect.

## Cartoon examples

An important step in analyzing whether there is cause and effect is to describe the purported cause and purported effect. By using cartoons to illustrate what we might see, we can examine that process, too. In what follows, I'll often say "the cause" or "the effect" when it should be clear that what I mean is "the purported cause" or "the purported effect" as described by claims.

The first issue we must resolve is how we can, even in theory, establish cause and effect. What can we do except check that each of the necessary conditions for the usual notion hold? In practice those conditions serve as sufficient, and I'll proceed as if they do. In the following sections I'll expand on many of the issues raised here.

*Example 13*  The cat made Spot run away.

*Purported cause*  What is the cause? It's not just the existence of the cat. How can we describe it with a claim? Let's take it to be "A cat meowed close to Spot."

*Purported effect*  Spot ran away.

*Cause and effect each happened*  The effect is clearly true. The cause is highly plausible, for almost all things that meow are cats, at least where Dick would be walking Spot.

*It's (nearly) impossible for the cause to be true and effect false*
This is not clear. We have to establish the normal conditions. Some appeal to Spot's usual behavior must be made: In the past he's always chased a cat that meowed or made its presence known to him if it was close to him. That is, Spot normally chases cats, given the opportunity.

Even granting that, how shall we make these normal conditions precise? What is "given the opportunity"? We have no reason to believe he'll chase just any cat anywhere at any time at any distance from him. We do not know those normal conditions. And being unable to specify them, we can at best say that it is highly unlikely in this situation that the cat could meow and Spot not chase it.

*The cause makes a difference*  Would Spot have run away even if the cat had not meowed near him? It would seem that under the normal conditions of a walk with Dick he would not, since Dick is holding the leash loosely, not prepared for Spot to run away at any moment, and he seems surprised. But would Spot have chased the cat even if it had not meowed? Perhaps yes, if he had been aware of it.

So let's revise the cause to: "Spot wasn't aware a cat was near him, and the cat meowed close to Spot." Now we can reasonably believe that the cause made a difference.

*Cause precedes effect*  Yes.

*There is no common cause*  Perhaps the cat was hit by a meat truck and lots of meat fell out, and Spot ran away for that? No, Spot wouldn't have barked. Nor would he have growled.

Perhaps the cat is a hapless bystander in a fight between dogs, one of whom is Spot's friend. We do not know if this is the case. So it is possible that there is a common cause, but it seems unlikely.

*Evaluation*  The relation between purported cause and purported effect cannot be established to be valid, in part because we cannot specify the normal conditions with good precision. Further, there may be a common cause. Nonetheless, we have good reason to believe the original claim on the revised interpretation that the cause is "Spot wasn't aware a cat was near him, and the cat meowed close to Spot."

These are the steps we should go through to establish a causal claim. If we can show that one of them fails, though, there's no need to check all the others.

*Example 14* The falling apple knocked Dick unconscious.

*Purported cause* An apple fell and hit Dick.

*Purported effect* Dick is unconscious.

*Cause and effect each happened* Yes.

*It's (nearly) impossible for the cause to be true and effect false*
We're not sure what the normal conditions are (what height the apple fell from, the speed of the apple, the weight of the apple, the thickness of Dick's skull). But we suspect that this is the case. We know that in the past hard knocks to people's heads are sometimes followed by unconsciousness. This is a generalization.

We have reason to believe the inference is valid, relative to unspecified normal conditions. That is, we suspect that we could make this into a valid inference if we had enough information.

*The cause makes a difference* This seems very likely. In our experience people do not just become unconscious while reading a book under a tree.

*Cause precedes effect* Yes.

*There is no common cause* There seems to be no common cause. The wind blowing may have caused the apple to fall, but that is not a common cause: The inference "The wind blew therefore Dick is unconscious" needs "An apple fell and hit Dick" as premise in order to be strong or valid. *Tracing the cause back farther is not the same as exhibiting a common cause.*

*Evaluation* This is about as plausible a causal claim as we are likely to encounter in our daily lives. Some might say it is implausible, because an apple is not heavy enough to knock someone unconscious. But then we ask what else caused Dick to become unconscious immediately after being hit on the head by an apple? There seems to be no candidate.

But isn't this shifting the burden of proof? Don't we have to show that there is a cause, rather than assuming there is one? No, this isn't to assume that everything has a cause, which we'll consider below. In this case we know from experience that there must be some unusual occurrence for someone to become unconscious, and we are looking for that condition to label as "cause."

*Example 15* The wasps chased Professor Zzzyzzx because he hit their nest.

*Purported cause* Professor Zzzyzzx hit a wasps' nest.

*Purported effect* The wasps chased Professor Zzzyzzx.

*Cause and effect each happened* It appears that this is a wasps' nest.

*It's (nearly) impossible for the cause to be true and effect false* We have enough examples of people hitting wasps' nests and being chased by wasps that we believe it is highly unlikely to hit a wasps' nest and not be chased by wasps. Here the normal conditions are: Wasps are in the nest, and no precautions were taken to ensure the wasps wouldn't chase the person hitting the nest (smoke, etc.). Note the generalization needed here.

*The cause makes a difference* It seems very unlikely that the wasps would have chased Professor Zzzyzzx had he not hit their nest. At least that seems so under the normal conditions: Professor Zzzyzzx wasn't touching or disturbing their nest or approaching it very closely other than to hit it.

*Cause precedes effect* Yes.

*There is no common cause* None is apparent.

*Evaluation* This is a highly plausible causal claim.

34   *Cause and Effect*

*Example 16*   Dick got burned because he put too much lighter fluid on the barbecue.

*Purported cause*   Dick put too much lighter fluid on the barbecue.
*Purported effect*   Dick got burned.

*Cause and effect each happened*   First we must establish that the liquid Dick is spraying is lighter fluid. Premises: Dick is spraying the fluid onto barbecue coals before dropping a match on the coals. You don't put water on coals before lighting them. It is extraordinarily rare to see gasoline or kerosene put into a container like the one Dick is holding. No other flammable fluids are likely to be put on coals. This fluid flamed up when the match hit it. This fluid is flammable.

So we have good reason to believe the cause is true, if by "too much" we mean more than the amount typically specified on the label of lighter fluid containers.

The effect is clearly true.

*It's (nearly) impossible for the cause to be true and effect false*
We depend on past experience here: Large amounts of lighter fluid that are lit typically flame up. People who are close to those flames will get burnt. These are generalizations.

But then we have to ask whether it is part of the normal conditions for lighting a barbecue to stand as close to it as Dick is standing when he puts the match on the coals. That does not seem to be normal. So we have to add to the description of the cause: "Dick is standing very close to the barbecue."

*The cause makes a difference*   Could it be that Dick would get burned if he had not put so much lighter fluid on the coals? We rely again on our past experience that normal amounts of lighter fluid do not flame up enough to reach a person standing near the grill.

*Cause precedes effect*   Yes.

*There is no common cause*   None is apparent.

*Evaluation* This is as plausible a causal claim as we are likely to get in ordinary experience, even though our justification of that relies on experience that we cannot be sure is universal.

*Example 17* Suzy failed because she stayed up late dancing.

*Purported cause* Suzy stayed up late dancing March 4th.

*Purported effect* Suzy failed the exam March 5th.

*Cause and effect each happened* The cause appears to be true: there are stars showing in the window, so Suzy and Tom are dancing at night. The next day Suzy and Tom appear tired, so it seems likely they were up late. The effect appears true, assuming these are grades for that exam

*It's (nearly) impossible for the cause to be true and effect false* This does not seem to be the case. After all, Tom stayed up late and passed the exam. However, when Suzy is tired she may perform worse on exams than Tom does when he's tired. But perhaps Tom studied more than Suzy.

We don't know the normal conditions. Did Suzy study before the exam? How much sleep did Suzy get the previous night?

*The cause makes a difference*  Would Suzy have failed if she hadn't stayed up late partying? We don't know.

*Cause precedes effect*  Yes.

*There is no common cause*  We don't know. Perhaps Suzy has the disposition both to party before exams and not to study.

*Evaluation*  We do not have enough information to judge this claim, so we suspend judgment. To assume otherwise is to make the most common mistake in causal reasoning.

---

***Post hoc ergo propter hoc***  It's a mistake to argue that there is cause and effect just because one claim became true after another.

---

*Example 18*  Dick crashed because of the turtle.

*Purported cause*  The turtle is crossing the road.

*Purported effect*  Dick crashed.

*Cause and effect each happened*  Yes.

*It's (nearly) impossible for the cause to be true and effect false*
It's easy to imagine lots of ways in which the cause could be true and effect false. For instance, Dick simply could have run over the turtle. Or Dick could have swerved and missed the turtle. Or Dick could have stopped in front of the turtle.

But were these possibilities? What are the normal conditions? Dick is going fast. But is that normal or part of the cause? Is Dick's normal reaction in very little time to swerve to avoid hitting an animal on the road? Dick seems to be paying attention and can see far down the road. Is that normal or is the negation of that a joint cause? After all, turtles don't just dart out into the road. But perhaps turtles are too low for someone driving a car to see easily.

*The cause makes a difference*   Yes, under normal conditions, Dick would not have crashed if he had not swerved to avoid the turtle.

*Cause precedes effect*   Yes.

*There is no common cause*   None is apparent. We might want to trace the cause back to the reason for the turtle being on the road ("Why did the turtle cross the road?"), but we don't know that, and the normal conditions would multiply too much.

*Evaluation*   Though we may be inclined to believe the causal claim is true, it seems more likely that there are other claims that we do not want to ascribe to the normal conditions that should be considered part of the description of the cause.

*Summary of issues discussed in the examples*
• We're often willing to accept a causal claim from ordinary life even though we can establish only that the relation between purported cause and purported is strong rather than valid. Often that's because it's impossible to specify normal conditions that would guarantee the validity of the inference. It's not that it is very difficult to establish that the relation of purported cause to purported effect is valid. Rather, in many cases against the background of any normal conditions we are likely to have any reason to believe, the relation will at best be strong.

• It is much easier to show there isn't cause and effect than to show there is. The difficulty is with the normal conditions.

• Reasoning about cause and effect is as hard as judging whether an argument is strong when the unstated premises to make it strong are so obvious that we can't even think of saying them, and when we do say them it seems stupid to point them out. Indeed, typically there are so many normal conditions that it would be impossible to state them all.

• Tracing the cause backwards is not the same as finding a common cause.

• Generalizations are often required as premises in causal inferences, though they may not be based on much experience.

These examples suggest three questions we'll consider next:

*38    Cause and Effect*

What is the role of generalizations in causal inferences?

Can a causal inference that is strong and not valid be good?

How do we distinguish normal conditions from cause?

## Generalizations in causal reasoning

*Example 19*  Zoe: Every time I wash my car, it rains within twelve hours.
Suzy: Well, don't wash your car today. I want my picnic to be fun.

*Analysis*  Behind Suzy's comment is a general causal claim: "Zoe's washing her car causes it to rain." We just laugh. Of course there's no connection.

But suppose it was always clear and forecast sunny for the next two days when Zoe washed her car. And it always rained within six hours. And this happened thirty times over two years. We'd have pretty good evidence for Zoe's claim.

Still we'd be suspicious. Constant conjunction isn't enough to convince us that if the cause weren't true, the effect wouldn't be true —the conjunction might be coincidence or the result of a common cause. My pulse is evidence that I'm breathing, occurring always in conjunction with it, and if I had no pulse I would have no breath. But my having a pulse is not the cause of my breathing.

Nor is constant conjunction necessary. Suppose in Example 18 we knew more about the speed Dick was driving and his feelings about killing small animals by accident, and we were willing say that the turtle caused the crash. How often have we seen turtles crossing a road in front of Dick speeding down the highway in his car? Maybe we should ask how many times we have seen animals of any sort crossing in front of cars speeding down a highway in this kind of situation. A lot depends on what is meant by "this kind of situation." Even then, we may have witnessed no such cases, yet be willing to classify the crash as an example of cause and effect. Similarly, an astronomer who sights a completely novel kind of phenomenon in the sky will rightfully look for the cause of that, knowing well that there is no correlation she can invoke since the event is of an entirely new kind.

We want a general principle or theory that connects cause and effect. The constant conjunctions give us motive to find one, but until we do we are apt to dismiss the causal claim as *post hoc* reasoning.

Compare: "The gas produced by cows' digestion causes the atmosphere to warm up." What evidence do we have? There are more cows now than ever before and it's getting warmer. Pretty slim evidence. But we have a theory. Scientists estimate the amount of methane produced by cows, calculate how this traps heat in the atmosphere, and say that the gas from cows is one of many causes that the atmosphere is warming.

*Example 20* The sun causes the earth to warm up.

*Analysis* This is a general causal claim. To be good, each of the following must be a good causal inference:

(*) The sun shined on the earth at this time, therefore the earth in this area where the sun was shining warmed up at this later time.

A generalization is needed to show that (*) is a good causal inference. At the least, it would seem, we would have to cull from our experience that when the sun shone it was always followed by the earth warming up at that place.

But we do not have that. Some days it gets colder after sunrise, especially in the winter in northern latitudes if the sun burns off a cloud cover that is holding in the heat. We use a more general theory about how the sun's rays affect the atmosphere, which is certainly not just an association of repeated instances of the sun shining and the area around us warming up. The theory has to account for why sometimes the sun shines and the area around us doesn't warm up. If the theory, which is just a collection of claims we believe are true, has as consequence each claim needed for (*) that "ought to be true," then we believe we have established this example. General claims, perhaps established by generalizations, play the same role in establishing causal inferences as in establishing other kinds of inferences.[10]

*Example 21* Any time I end up in an enclosed space with a cat for more than about fifteen minutes my eyes start itching, I start sneezing, and I get an asthma attack. Cats make me sneeze and give me asthma attacks.

*Analysis* This is a general causal claim, with an instance of cause "I was in an enclosed space with a cat for more than about fifteen minutes" and effect "I started sneezing and got an asthma attack." The general claim of the example is true. But does the causal claim follow?

---

[10] See *The Fundamentals of Argument Analysis* in this series.

The correlation gives us reason to believe that it is highly unlikely the cause could be true and effect false. All the other conditions for cause and effect seem to hold, except for the question whether the cause makes a difference. If there had been no cat, would I have sneezed and had an asthma attack anyway? Here the normal conditions are crucial, for I do sometimes sneeze and have an asthma attack in an enclosed space in which there is no cat. We have to assume there was no cat or horse hair in the room, or mold, or house dust, or alfalfa, or else expand the cause to exclude the presence of those. Then the cause is as well established as the correlation.

*Example 22* Gravity caused the apple to fall (in Example 14).

*Analysis* What is the purported cause here? Isn't it just the existence of a thing, gravity, that is invoked as the cause?

Gravity is a name given to the notion of a causal power inhering in things that makes them attract other things.[11] It is replaced in physics by generalizations, formulas which, given values for objects of various

---

[11] William A. Wallace describes the state of the controversy regarding the nature of causes at the end of the nineteenth century in his *Causality and Scientific Explanation*, p. 159:

> Chief among the topics of controversy was whether causes should be viewed uniquely as events, as discernible only at the phenomenal level, and as expressible in some way in terms of laws regulating phenomena, or whether they should be taken more in the sense of forces, internal mechanisms, latent configurations, or hidden processes that serve to explain phenomenal occurrences. Implicit in the answers to these questions were actually two meanings of causality, neither of which was clearly accepted by the scientific community, nor was the difference between them sufficiently realized. The first may be characterized as a weak or minimal meaning, usually associated with the term "causation", which stresses the phenomenal or eventlike character of cause and effect and therefore holds that causes themselves can be discovered and known; whereas the relation between cause and effect, not being an observable event or occurrence, is best interpreted as a psychological or conceptual projection into reality. The second meaning, on the other hand, is the strong or maximal sense usually associated with the use of the term "cause" in ordinary language. This stresses knowledge of the connection between cause and effect, generally in terms of some agency or efficacy discerned in the effect's production, and seeing this as lodged in some way in the mechanism and conditions necessary for such production.

sizes and mass and distance from each other, yield values for the acceleration of objects. This led Bertrand Russell to argue that modern science dispenses with causes:

> Laws of probable sequence, though useful in daily life and in the infancy of a science, tend to be displaced by quite different laws as soon as a science is successful. The law of gravitation will illustrate what occurs in any advanced science. In the motions of mutually gravitating bodies, there is nothing that can be called a cause, and nothing that can be called an effect; there is merely a formula. Certain differential equations can be found, which hold at every instant for every particle of the system, and which, given the configurations at two instants, render the configuration at any other earlier or later instant theoretically calculable. That is to say, the configuration at any instant is a function of that instant and the configurations at two given instants. This statement holds throughout physics, and not only in the special case of gravitation. But there is nothing that could be properly called "cause" and nothing that could be properly called "effect" in such a system.[12]

All we need are functional relationships.

What Russell points out is that modern science has (often) dispensed with causes in the sense of powers inhering in objects.[13]

---

[12] "On the notion of cause," p. 194.

[13] Russell was not the first to argue for the elimination of causes in science. Newton had already argued against ultimate causes (see Appendix 1 on causes and reasons). Auguste Comte in his *Cours de philosophie positive* (1830–1842) also said that causes were not needed in science, as detailed in William Wallace's *Causality and Scientific Explanation*, vol. 2, pp. 85–86:

> The law is this: that each of our leading conceptions—each branch of our knowledge—passes successively through three different stages: the theological, or fictitious; the metaphysical, or abstract; and the scientific, or positive. In other words, the human mind, by its nature, employs in its progress three methods of philosophizing whose characteristics are essentially different and even radically opposed: first the theological method, then the metaphysical, and finally the positive. . . . The first is the necessary point of departure of the human understanding and the third is its fixed and definitive state; the second is merely a state of transition. [Comte, *Cours*, vol. 1, pp. 3–4]

When at the theological stage, the human mind, supposing that all phenomena are produced by the immediate action of supernatural beings, seeks "the essential nature of beings, the first and final causes of all effects." At the metaphysical stage the mind replaces supernatural agents with "abstract forces" and other

*42   Cause and Effect*

But that isn't to say that science has or can dispense with cause and effect reasoning. In the case at hand, the functional relation serves as a generalization that supports the causal inference:

> The apple was this far from the center of the earth and this far from the surface of the earth and was not connected to any other object, therefore the apple fell.

We avoid invoking a thing or a power inherent in a thing as the cause just as the scientist does. But science does not dispense with the notion of cause in that it does not dispense with certain kinds of inferences.

## Can a strong causal inference be good?

Some say that it is our ignorance that leads us to accept strong causal inferences as good. But the belief that we cannot specify the normal conditions for what "really" is a valid causal inference is no more than an article of faith, a restatement of the requirement that good causal inferences must be valid, not an explanation for why they should be.[14]

---

entities that are "personified abstractions," regarding these as capable of producing all phenomena. It is only at the last, or positive, stage, that the mind gives over "the vain search after absolute notions, the origins and destination of the universe, and the deeper causes of phenomena, and applies itself . . . to the study of the laws—that is, their invariable relations of succession and resemblance." The knowledge of such laws is attained by observation and reasoning, and they provide true scientific explanations. As Comte expresses it:

> The explanation of facts, reduced to its real terms, will henceforth be only the relationships established between different particular phenomena and some general facts, the number of which tend to decrease more and more with the progress of science. [Comte, *Cours*, vol. 1, p. 5]

See also the quote by John Stuart Mill in footnote 25 below.

The notion of a causal power is still used by some. John R. Searle in "I married a computer," p. 34, says:

> I believe that there is no objection in principle to constructing an artificial hardware system that would duplicate the powers of the brain to cause consciousness using some chemistry different from neurons.

Richard Boyd in "Observations, explanatory power, and simplicity: toward a non-Humean account" presents and defends the view that causal powers and mechanisms are essential in the analysis of cause and effect.

[14] See G.E.M. Anscombe, "Causality and determination," p. 91. To assume that every good causal inference is valid is comparable to the assumption that

The only role that assumption plays in our reasoning is to goad us to try to specify better normal conditions that would tighten, if not make perfect, the correlation between the purported cause and effect.

It seems to me we have ample reason to accept some strong causal inferences as good. Whether you agree or not, putting the issue in these terms allows us to see the problem in the general context of inference analysis and not confuse the issue of whether the claims describing the cause and effect are true with whether there is a necessary connection between cause and effect if they are true, as Max Born apparently does:

> The first use of probability considerations in science was made by Gauss in his theory of experimental errors. . . . It has a direct bearing on the method of inference by induction which is the backbone of all human experience. I have said that in my opinion the significance of this method of science consists in the establishment of a code of rules which form the constitution of science itself. Now the curious situation arises that this code of rules, which ensures the possibility of scientific laws, in particular of the cause-effect relation, contains besides many other prescriptions those related to observational errors, a branch of the theory of probability. This shows that the conception of chance enters into the very first steps of scientific activity, in virtue of the fact that no observation is absolutely correct. I think chance is a more fundamental conception than causality; for whether in a concrete case a cause-effect relation holds or not can only be judged by applying the laws of chance to the observation.[15]

*Example 23* A bomb is connected to a Geiger counter. If the reading on the counter is sufficiently high, the bomb will go off. A piece of radium is placed next to the Geiger counter. One day the Geiger counter detects sufficient radiation and the bomb goes off. The radioactive decay of the radium caused the bomb to go off.

*Analysis* The purported cause is "The radium decayed radioactively to a particular level," or perhaps better, "The radium emitted $n$ number of atomic particles at time $t$." In that case, there is a necessary relation of cause to effect: shortly after the radium emitted that number of particles, the Geiger counter read sufficiently high, and then the bomb exploded.[16]

---

everything is physically determined, it's just that the computations are too complex for us to carry out.

[15] *Natural Philosophy of Cause and Chance,* pp. 46–47. By "principle of causality" I take it he means the "law of causation" discussed below.

The problem, though, is if we wish to trace the cause back to "Someone put the radium near the Geiger counter and bomb mechanism."[17] The inference from that to the effect "The bomb went off " is only strong: statistically it is very likely that enough particles will be emitted by the radium at some moment to trigger the bomb, but it is not necessary according to current theories of physics. Thus, there is cause and effect if we allow strong causal inferences to be good. But if a causal inference must be valid to be good, then this example (and many others from physics) would not be classified as cause and effect.[18]

---

[16] This example is sometimes used to suggest that the universe is not determined: knowing the initial conditions of the bomb, Geiger counter, radium, etc., it is not possible, even in theory, to predict when or if the bomb will ever go off, as in G. E. M. Anscombe, "Causality and determination", pp. 100ff.

[17] This is what Anscombe does not consider. See the discussion in Example 33 below from H. L. A. Hart and Tony Honoré of tracing the cause back to the intervention of a human agent.

[18] See Appendix 3 on objective chance for another view on probability in reasoning about cause and effect. Friedrich Waismann in "Verifiability" discusses this issue in relation to Kant's ideas (see footnote 4 above):

> What matters is, not whether quantum mechanics draws a true picture of reality, but only whether it draws a *permissible* one. About that there can be little doubt. Kant was of the opinion that if there was no such thing as causality science would simply break down. Now the important thing that has emerged is the *possibility* of constructing a theory along different lines, the *legitimacy* of departing from causality, while science has not died or committed suicide on that account. This suffices to disown any claim on the part of Kant to regard causality as an *indispensable* form of our knowledge of the world. Had he been right, we could not even *entertain* such views as physicists do to-day; to give up causality, even in part, would mean to rob us of the very condition of gaining knowledge; which could end in one result only, in complete confusion. But that is not so. Though causality has been severely limited, quantum mechanics is a useful tool. Kant did not foresee the possible forms of physical laws; by laying too much stress on the scheme of causality, by claiming for it an *a priori* status, he unduly narrowed the field of research. p. 50
>
> But leaving quantum mechanics and turning to the common world of sense, I still fail to see any ground for accepting Kant's position. True, in order to orient ourselves in the world we must presuppose that there is some sort of order in it so that we may

## The normal conditions
It's not clear by what standard we distinguish normal conditions from the cause.

*Example 24*   Dick:  Wasn't it awful what happened to old Mr. Green?
   Zoe:  You mean those tree trimmers who dropped a huge branch on him and killed him?
   Dick:  You only got half the story.  He had a heart attack in his car and pulled over to the side.  He got out to get help and was lying on the pavement when the branch hit him—he would have died anyway.
   *Analysis*   The purported cause is "The tree trimmers dropped a huge branch on Mr. Green," and the purported effect is "Mr. Green died." Both are true, and it's clear that it's impossible for the cause to be true and effect false.  But did the cause make a difference?
   Mr. Green would have died anyway.  So no, the tree branch falling on him wouldn't have made a difference.  But that's wrong: his heart attack was not sufficient for his dying at that moment in that way.  That is, we are not willing to say the following are equivalent:

Mr. Green died at the moment the tree branch hit his skull.
Mr. Green died shortly after he had a heart attack.
Mr. Green died.

*The description of what happened matters for causal analyses because different claims result in different inferences.*
   The issue is why to take or prefer one claim to another as a "description of what happened."  Mrs. Green surely prefers the first claim as describing the effect in order to collect from the tree trimmers.

*Example 25*   (Socrates is sitting in the chamber where he is to drink his potion of hemlock.)  The cause of Socrates sitting in the chamber is that his muscles are contracted appropriately.
   *Analysis*   In the *Phaedo* Socrates is rehearsing some of the faulty ideas of cause and effect he held earlier in his life.  He gives the following example:

---

anticipate the course of events and act accordingly.  What I fail to see, however, is why this order should be a strictly *causal* one. Suppose, for the sake of argument, that the objects around us were, *on the average*, to display an orderly behaviour, then the world may still be a liveable place.   p. 51

> The reason why I am lying here now is that my body is composed of bones and sinews, and that the bones are rigid and separated at the joints, but the sinews are capable of contraction and relaxation, and form an envelope for the bones with the help of flesh and skin, the latter holding all together, and since the bones move freely in their joints the sinews by relaxing and contracting enable me somehow to bend my limbs, and that is the cause of my sitting here in a bent position.[19]

He continues by showing that this is incorrect:

> Or again, if he tried to account in the same way for my conversing with you, adducing causes such as sound and air and hearing and a thousand others, and never troubled to mention the real reasons, which are that since Athens has thought it better to condemn me, therefore I for my part have thought it better to sit here, and more right to stay and submit to whatever penalty she orders. Because *by dog,* I fancy that these sinews and bones would have been in the neighborhood of Megara or Boeotia long ago—impelled by a conviction of what is best!—if I did not think it was more right and honorable to submit to whatever penalty my country orders rather than take to my heels and run away. But to call things like that causes is too absurd. If it were said that without such bones and sinews and all the rest of them I should not be able to do what I think is right, it would be true. But to say that it is because of them that I do what I am doing, and not through choice of what is best—although my actions are controlled by my mind—would be a very lax and inaccurate form of expression. Fancy being unable to distinguish between the cause of a thing and the condition without which it could not be a cause![20]

A cause must make a difference: If the cause had not happened, the effect would not be. The cause is a condition ***sine qua non*** (without which not). But as Socrates points out, not every condition *sine qua non* can be classified as a cause. The claims about sinews and bones are part of the normal conditions for Socrates sitting in the chamber.

Two questions arise from the this example. First, is it possible for there to be a cause which is not *sine qua non*? We'll consider that in Example 28. But since we have not required causal inferences to be valid, we must understand *sine qua non* in the sense of a condition

---

[19] Plato's *Phaedo*, 98c–d, as translated by Hugh Tredennick.
[20] Ibid, 98d–99b, italics added. See Appendix 1 of this essay concerning Plato's use of "reason" and "cause."

needed for perhaps only a strong inference. As there are many inferences to any conclusion, what may be *sine qua non* for one may not be for another.

The second question is more difficult. How are we to distinguish among all the conditions *sine qua non* that one or those that we are to call "cause(s)"? The problem isn't what we can add as unstated premises. The problem is that the claim is already there in the inference, picked out as one that isn't "normal."

*Example 26* A person eats a dish that is spoiled, and then he dies. The cause of his death is his eating the dish.
*Analysis* John Stuart Mill discusses this example:

> It is seldom, if ever, between a consequent and a single antecedent, that this invariable sequence subsists. It is usually between a consequent and the sum of several antecedents; the concurrence of all of them being requisite to produce, that is, to be certain of being followed by, the consequent. In such cases it is very common to single out one only of the antecedents under the denomination of Cause, calling the others merely Conditions. Thus, if a person eats of a particular dish, and dies in consequence, that is, would not have died if he had not eaten of it, people would be apt to say that eating that dish was the cause of his death. There need not, however, be any invariable connection between eating of the dish and death; but there certainly is, among the circumstances which took place, some combination or other on which death is invariably consequent: as, for instance, the act of eating the dish, combined with a particular bodily constitution, a particular state of present health, and perhaps even a certain state of the atmosphere; the whole of which circumstances perhaps constituted in this particular case the *conditions* of the phenomenon, or, in other words, the set of antecedents which determined it, and but for which it would not have happened. The real Cause, is the whole of these antecedents; and we have, philosophically speaking, no right to give the name of cause to one of them, exclusively of the others.[21]

H. L. A. Hart and Tony Honoré disagree:

> When we assert that *A*'s blow made *B*'s nose bleed or *A*'s exposure of the wax to the flame caused it to melt, the general knowledge used here is knowledge of the familiar way to produce, by manipulating things, certain types of change which do not normally occur without

---

[21] *A System of Logic,* Book III, Chapter V, section 3 (p. 237).

our intervention. If formulated they are broadly framed generalizations, more like recipes, in which we assert that doing one thing will "under normal conditions" produce another, than statements of "invariable sequence" between a complex set of specified conditions and an event of the given kind. Mill's description of common sense "selecting" the cause from such a set of conditions is a *suggestio falsi* so far as these simple causal statements are concerned for, though we may gradually come to know more and more of the conditions required for our interventions to be successful, we do not "select" from them the one we treat as a cause. Our intervention is regarded as the cause from the start before we learn more than a few of the other necessary conditions. We simply continue to call it the cause when we know more.

It is, moreover, a marked feature of these simple causal statements that we do not regard them as asserted unjustifiably or without warrant in a particular case if the person who makes them cannot specify any considerable number of the further required conditions.[22]

We note with a claim some unusual state of the world and call it "the cause," even though we have not reflected on what exactly was normal when that occurred (the time of which it is true). What is normal is not only not usually worth remarking, it is not even noted by us, and may be difficult if not impossible for us to state. This seems not only what we do, but what we must do if we are to use the notion of cause and effect in reasoning about our ordinary lives, as the cartoon examples show.

As Michael Scriven says,

> When we identify a cause, we are doing so on the basis of contextual inference as to what *type* of cause is sought, and with what degree of precision it must be described in order to be completely described, with what is known about the surrounding circumstances (from which we infer what counts as an abnormal or notable circumstance), etc.[23]

And Norwood Russell Hanson says:

> The primary reason for referring to the cause of *x* is to explain *x*. There are as many causes of *x* as there are explanations of *x*. Consider how the cause of death might have been set out by a physician as "multiple haemorrhage", by a barrister as "negligence on the part of the driver", by a carriage builder as a "defect in the brakeblock construction", by a civic planner as "the presence of tall shrubbery at that turning". . . .

---

[22] *Causation in the Law*, p. 31.
[23] "Explanations, predictions, and laws," p. 215.

## Cause and Effect 49

Nothing can be explained to us if we do not help. We have had an explanation of *x* only when we can set it into an interlocking pattern of concepts about other things, *y* and *z*. [24]

To the extent that what we consider to be normal depends on our experience and expectations, what we classify as cause or as a normal condition is subjective.[25] It is intimately connected to what we would accept as an adequate explanation, as discussed in "Explanations." Still, this is not to deny that there might be an objective distinction and what we give are just subjective evaluations of that.

*Example 27* Harry works in a laboratory where there's not supposed to be any oxygen. The materials are highly flammable and he has to wear breathing gear. He was joking around with a friend and struck a match, knowing it wouldn't ignite. It seems there was a leak in his face mask. Harry striking the match caused the match to burn.

*Analysis* The purported cause is "Harry struck the match," and the purported effect is "The match burned." The normal conditions here don't include "Oxygen is in the laboratory." That along with Harry striking the match caused the match to burn. *There may be several claims we want to say jointly describe the cause*: Oxygen

---

[24] *Patterns of Discovery*, p. 54.

[25] Mill attempts a different criterion for picking out the cause from the normal conditions. Continuing the quote on p. 47 above, he says:

> What, in the case we have supposed, disguises the incorrectness of the expression, is this: that the various conditions, except the single one of eating the food, were not *events* (that is, instantaneous changes, or successions of instantaneous changes) but *states*, possessing more or less of permanency; and might therefore have preceded the effect by an indefinite length of duration, for want of the event which was requisite to complete the required concurrence of conditions: while as soon as that event, eating the food, occurs, no other cause is waited for, but the effect begins immediately to take place: and hence the appearance is presented of a more immediate and close connection between the effect and that one antecedent, than between the effect and the remaining conditions. But though we may think proper to give the name of cause to that one condition, the fulfillment of which completes the tale, and brings about the effect without further delay; this condition has really no closer relation to the effect than any of the other conditions has.

But this emphasis on change does not always seem warranted; see Example 7.

was in the laboratory; Harry carried matches into the laboratory with him; Harry struck the match. The rest can be relegated to the normal conditions.

---

*A cause*  When there is more than one claim which we wish to distinguish from the normal conditions as cause, we say that each describes *a cause*. That is, each of A and B is a cause of C if "A and B, therefore C" is a causal inference, and neither "A therefore C" nor "B therefore C" is a good causal inference, and neither A nor B is classified as normal.[26]

---

*Example 28*  (Ralph and Harriet are the only members of a firing squad and have been instructed to shoot Rodolfo. Ralph aims at Rodolfo's head, and Harriet aims at his heart. They both have very large caliber guns. On signal, they shoot, and then at exactly the same moment Ralph's bullet obliterates Rodolfo's head and Harriet's bullet obliterates Rodolfo's heart.)

Ralph caused Rodolfo to die.

*Analysis*  I take the purported cause to be described by "Ralph shot Rodolfo in the head."

Here is a case where all the conditions for there to be cause and effect appear to be satisfied with the possible exception that the cause made a difference. The purported cause is not a condition *sine qua non* of the effect "Rodolfo died" since that would have happened anyway from Harriet's bullet. There seem to be two equally good candidates for the cause. We can say that an effect is *causally overdetermined* when there are two (or more) non-equivalent claims true of the same time such that each independently qualifies as describing the cause relative to all normal conditions that preceded the effect except that the cause makes a difference, and that condition would hold were the other claim not true.

Still, both Ralph and Harriet are shooting due to a command, and it does not seem to be tracing the cause too far back to say that the cause of Rodolfo's death is that they were commanded to shoot him. Still, if this example does not sufficiently call into question whether a cause

---

[26] Igal Kvart, "Cause and some positive causal impact," pp. 408–409, distinguishes between "the" cause and "a" cause by saying the former is context dependent and relative to our interests, while the latter is objective, interpreting the latter in terms of objective chance; see Appendix 3 below.

must be a condition *sine qua non*, modify it to: Two people shoot a third simultaneously where neither is aware of the other and each has different motives for shooting.

Perhaps there, too, we could trace the cause back farther in time, ascribing the cause to what the person who was shot did or said or acted over a period of time. On the face of it there is nothing to preclude this approach, but it will always smack of being *ad hoc*.

Hart and Honoré suggest that we call each of A and B "the cause":

> The difficulty in such cases can be simply stated. Two sufficient causes of an event of a given kind are present and, however fine-grained or precise we make our description of the event, we can find nothing which shows that it was the outcome of the causal process initiated by the one rather than the other. It is perfectly intelligible that in these circumstances a legal system should treat each as the cause rather than neither, as a *sine qua non* test would require.[27]

That is, for the exposition of the law we need the idea that a cause may be merely sufficient for an effect that has happened, and that there are genuine cases of overdetermination.[28]

Consonant with their observation is to say that the two claims A and B jointly describe the way the world was at a certain time, so together they describe the cause. We recognize that each of A and B is too narrow a description of the world.[29] *Causal overdetermination is just a result of too narrow descriptions of the world being proffered as the cause.*

*Example 29* Maria: Fear of getting fired causes me to get to work on time.

 *Analysis* What is fear? The cause is "Maria is afraid of getting

---

[27] *Causation in the Law*, pp. 124–125.
[28] *Causation in the Law*, p. xlii. The "law" here means legal rulings, etc.
[29] J. L. Mackie in *The Cement of the Universe*, p. 47, says:

> It is such clusters [of events] that we can confidently take as causing these effects. "But which item in the cluster really caused (or 'brought about') this effect (or 'result')?" is a sensible question in so far as it asks for a discrimination between the alternative "causes" by way of the filling in of a more detailed account. But if no more detailed account would provide the desired discrimination, this question has no answer.

What he calls a "cluster of events" is just a way the world was that made those several claims true at the same time, which we describe with several claims.

fired." The effect is "Maria always gets to work on time." This is to interpret the example as a general causal claim, each workday being an instance of it.

Is it possible for Maria to be afraid of getting fired and still not get to work on time? Certainly, but not, perhaps, under normal conditions: Maria sets her alarm; the electricity doesn't go off; there isn't bad weather; Maria doesn't oversleep; . . . . But doesn't the causal claim mean that it's because she's afraid that Maria makes sure these claims will be true, or that she'll get to work even if one or more is false? She doesn't let herself oversleep because of her fear.

In that case how can we judge whether what Maria said is true? It's easy to think of cases where the cause is true and effect false. So we have to add normal conditions. But that Maria gets to work regardless of conditions that aren't normal is what makes her consider her fear to be the cause.

Subjective causes are often attributed because of some sense that we control what we do. They are often too vague for us to classify as true or false. We suspend judgment here.

*Example 30* The cause of the earth rotating now is that it was rotating earlier.

*Analysis* The rotation of the earth at an earlier time is part of the normal conditions for the earth rotating. Physics elevates that observation into a law, the law of inertia. If we were to call the previous rotation of the earth the/a cause of the earth rotating now, it seems we should also label my car existing a few minutes ago a cause of its existing now. We assume that it is normal for objects to persist in time, a law of inertia of objects.[30]

*Example 31* The cause of Dick waking up was that Spot was barking and (Spot is a dog or Spot is not a dog).

*Analysis* One charge laid against using claims to describe causes and analyzing the relation of cause to effect in terms of inferences is that any logical complication of a cause that yields an extensionally equivalent claim would also count as a cause.

In this case we have that "Spot was barking" is truth-functionally equivalent to "Spot was barking and (Spot is a dog or Spot is not a dog)." On what grounds do we distinguish the first and not the second as the cause?

---

[30] Compare J. L. Mackie, *The Cement of the Universe*, pp. 156–157.

We already have that "Spot was barking, therefore Dick woke up" is a good causal inference. The additional claim "Spot is a dog or Spot is not a dog" is an irrelevant premise (since it is part of a conjunction, it can be viewed as a separate premise). Here "irrelevant" has the precise meaning we have in inference analysis: The premise is not needed to make the inference good.[31]

***Causal relevance*** If $\Sigma$ is a collection of claims, and both "$\Sigma$ therefore B" and "$\Sigma$ and A therefore B" are good causal inferences, where A is not equivalent to some conjunction of claims in $\Sigma$, then A is not a causal factor.

This is only a refinement of the requirement that the cause makes a difference by separating the claims involved.[32]

Similarly, "Spot barked and Dick woke up" is not the cause of Dick waking up. Nor do we need to resolve the role of abstract objects in cause and effect to see that "Spot barked and $2 + 2 = 4$" is not the cause of Dick waking up.

*Example 32* A lights a fire in the open, a mild breeze gets up and the fire spreads and consumes the adjoining house. A's action was the cause of the destruction of the house.

*Analysis* This example is from Hart and Honoré, who say:

> There are some cases where it seems natural to pursue connections "through" a sequence of independent events: A lights a fire in the open, a mild breeze gets up and the fire spreads and consumes the adjoining house. Here the breeze was independent of the man's action in lighting the fire, but in terms of the law's favourite though perhaps most misleading metaphor, it does not "break the chain of causation". It would be correct and natural to say that A's action was the cause of the destruction of the house. But if a man shoots at his wife intending to kill her, and she takes refuge in her parents' house where she is injured by a falling tile, though we may believe, on the strength of various general propositions, that if the man had not shot at his wife she would not have been injured (just as we may believe also, on the strength of general propositions, that if A had not

---

[31] Compare Hart and Honoré, p. 116. A reviewer suggested that "Spot is a dog or Spot is not a dog" is a normal condition *par excellence*.

[32] Kvart defines causal relevance in terms of objective probabilities; see Appendix 3 here on objective chance.

lit the fire the house would not have been destroyed), this would not justify the assertion that the man had caused his wife's injury either in a legal context or any other. It is obvious that, when we pursue causal connection in particular cases through a series of events, we have to take account, not only of generalizations which may inform us what sorts of events are necessary or sufficient conditions of the occurrence of others, but also of considerations of a quite different order, concerning the way in which generalizations may be combined and applied in particular cases.[33]

I would say rather that the normal conditions can include what happens after the cause as well as before. If a claim becomes true after the cause that is not part of what normally would follow that cause, we must view it as a joint cause or as an *intervening cause*. It seems that Hart and Honoré consider that a tile falling from a roof is not normal, so that is either a joint cause or the cause of the woman's injury.

It also seems they consider the breeze starting up to be normal or a *foreseeable consequence* of lighting the fire, so they trace the cause in that case to setting the fire, easily filling in the normal conditions. But such breezes may occur only 5% of the time yet we still want to say the fire-setter is liable, for a prudent person should not take a 5% risk if the consequences might be disastrous. That is, we might want to count the breeze as "normal" relative to assessing liability, even if it is not statistically normal. Assigning blame in the law and determining cause are distinct issues.

*Example 33* A contractor in northern Utah cuts a power cable. Two hundred and fifty miles to the south Geraldo loses power for three hours in his home. When the power comes back there is a power surge and the logic board on his computer is destroyed. The contractor caused the destruction of his computer.

*Analysis* The cause here isn't too far away in space and time from the effect, for we can fill in the normal conditions and show that the causal inference is good.

Others would say that this purported cause is close enough in space and time to the effect because there is a causal chain that links cutting of the cable to the destruction of Geraldo's computer. Little step by little step of clearly causal inferences establish the causal claim.

---

[33] *Causation in the Law*, pp. 12–13.

*Recourse to "causal chains" is nothing more than to say that, as in any inference where a premise is claimed to be irrelevant to the conclusion, further premises that are highly plausible can be added to the inference to make it strong or valid.* We can fill out the normal conditions; when we can, we say the purported cause was close enough in space and time to the effect.

What we count as cause may reflect also how complex an inference we are able or willing to comprehend. Even if normal conditions can be supplied, the inference may become too complex for us to accept as a good causal inference. So we break the inference into smaller ones, a "chain of cause and effect," labeling the larger inference a different kind of cause, one that is not "proximate."

The relation of cause to effect is not always considered transitive. The problem isn't, as with counterfactuals, that we don't know what strength of inference is expected or that the normal conditions are not compatible.[34] The problem seems to be that we have some intuitive limit as to how complex normal conditions can be; when we go from A caused B, and B caused C, and C caused D, to A caused D, the complexity of the normal conditions for the last may make it seem an intractable inference, even if each of the preceding inferences are good. Hence many writers distinguish between "the direct causal relation" and the "hereditary" or "ancestral causal relation."[35]

Hart and Honoré suggest that "a voluntary intervention is a limit past which consequences are not traced."[36] They are generally interested in how what we label as cause and how we trace a cause backwards are affected by the presence of human agents in the actions.

> Causal relations are not always "transitive": a cause of a cause is not *always* treated as the cause of the "effect", even when the cause of the cause is something more naturally thought of as a cause than a man's motive is. . . . By contrast, we do not hesitate to trace the cause back through even very abnormal occurrences if the sequence is deliberately produced by some human agent. . . . The causal relationship is sometimes transitive in other cases: notably when there is some other ground for thinking of the intermediate causes as analogous to "means" by which the earlier cause produces its effects.

---

[34] See the essay "Conditionals" in this volume.
[35] For example, Michael Tooley in *Causation: A Realist Approach.*
[36] *Causation in the Law*, p. 75, which is contained in a larger discussion on pp. 71–78.

This will be so when it is *well known* that a given effect is likely, by leading to the intermediate cause, to lead to the effect.[37]

Their analysis suggests a program to give criteria for tracing causes backwards in terms of the nature of the claims involved in causal inferences.

*Example 34* A motorist is speeding and is later involved in an accident. His speed, though, only brings him "to the scene of the accident at the same time that it was reached by the victim, though at the time of the accident the motorist was no longer speeding or the speeding then made no difference to the outcome."[38]

The motorist speeding was the cause of the accident.

*Analysis* Here is what Hart and Honoré say:

> The excessive speed merely serves to secure that the motorist is present at a given place at a time earlier than would have been the case had he not speeded, and that the risk of an accident occurring at that earlier time is no greater than it would have been had he arrived later.[39]

They conclude that the causal claim is false. But why should we believe that there is no greater risk arriving earlier than later? After all, the motorist was involved in the accident at the earlier time.

Rather, to set out the normal conditions starting with the motorist speeding is beyond our abilities. They would be of the sort that links a butterfly fluttering its wings in a rainforest last week to it raining in Chicago today. So everything is connected in one big causal chain. But then there would be no point in identifying any claim as describing a cause.

Hart and Honoré, discussing proximate causes, say that many lawyers seem happy to accept the criterion of *sine qua non* as a matter of fact in questions of cause and effect.[40] But many claim that the notion of proximate cause is too vague to be a matter of fact and is just a matter of policy. As Hart and Honoré say:

> According to the legal interpretation of this doctrine anyone who asks for the cause of an event faces an *embarras de choix* of literally cosmic proportions. From this predicament the inquirer can be

---

[37] Ibid., p. 43.
[38] Ibid., p. xxxviii.
[39] Ibid., p. xxxix.
[40] Ibid., Chapter IV.

rescued only by "arbitrary" practical principles of selection or limitation supplied by the law.[41]

**Summary of issues discussed in the examples**
- A cause must be *sine qua non*. But many claims satisfy that condition, and the problem is to distinguish cause from normal conditions. What we classify as cause or normal condition may be a subjective evaluation, relative to the context or our interests. This is not to deny that such evaluations can or should reflect an objective standard.

- Two or more claims may jointly describe the cause, either because neither is normal or sufficient, or because each of them rules out the other as being *sine qua non*. In that case we say that each is *a* cause.

- We can define a notion of causal relevance in terms of whether a claim is needed to ensure that an inference is a good causal one.

- Sometimes the link between cause and effect is established by a "causal chain," a series of causal inferences. However, transitivity of the causal relation is sometimes denied, in part because of the complexity of the normal conditions or the complexity of the inference evaluation.

Now let's turn to questions about the nature and metaphysics of cause and effect.

## Can cause and effect be simultaneous?
The usual notion of cause and effect includes a metaphysics of time and space. Time is essential in distinguishing cause from effect: the cause precedes the effect. But some examples have been proposed that are said to show that cause and effect can be simultaneous.

*Example 35* Xantippe's dying caused Socrates to be a widower.
 *Analysis* This is cited as an example where cause is simultaneous with the effect. But this isn't cause and effect; it's definition: Xantippe died if and only if "Socrates is a widower" is true, where the "if" is the

---

[41] Ibid., p. 110. William L. Prosser in *Handbook of the Law of Torts*, p. 252, says:

> "Proximate cause," or "legal cause," is the name given to the limitation which the courts have been compelled to place, as a practical necessity, upon the actor's responsibility for his conduct. The limitation is nearly always a matter of various considerations of policy which have nothing to do with the fact of causation.

truth-functional conditional, not an inference. Compare Calvin Coolidge, "When more and more people are thrown out of work, unemployment results."

*Example 36*  My hand moving causes the pencil to move.
  *Analysis*  Here is how Richard Taylor presents this example:

> Consider the relationship between one's hand and a pencil one is writing with. It is surely true that the motion of the pencil is caused by the motion of the hand and not vice versa. This, we can suppose, means in part that the motion of the hand is sufficient for the motion of the pencil under the circumstances assumed to obtain. Given those circumstances, however, the motion of the pencil is also sufficient for the motion of the hand, for under the circumstances assumed—that the fingers are grasping the pencil in a certain way, and so on—neither the hand nor the pencil can move without the other's moving with it. It logically follows, then, that under the conditions assumed to obtain the motion of either is also necessary for the motion of the other. It further appears that the motions are contemporaneous; the motion of neither is followed by the motion of the other. They move together. And in that case one cannot distinguish cause from effect by any consideration of which occurs first. ... There appears to be no temporal gap between cause and effect.[42]

Consider three illustrations of what might happen:

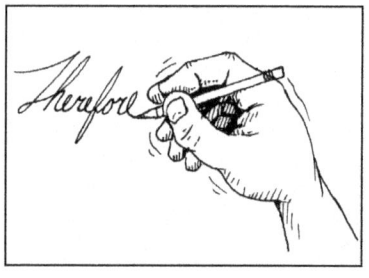
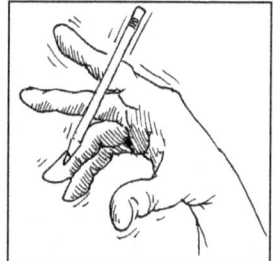

---

[42] "Causation," pp. 65–66.

The second picture shows that moving the hand is not sufficient to move the pencil *in that manner* (*writing*) *at that time*. The moment the hand moves I may choose to or accidentally release the pencil.

In the third picture the hand and pencil together are moved by another person's hand; a paralytic may have only partial control of her hand and ask a friend to help her while she speaks what she wants written. Here we would not say that the paralytic's hand moving caused the pencil to move.

Rather, in all these cases it is an antecedent to the hand and pencil moving that we should call the cause. For the first picture, it is "I will my hand to move while grasping the pencil." The normal conditions describe the normal functioning of my brain and hand and also include that nothing is impeding my hand. All we know of medical science now shows that there is brain activity that we can ascribe as either an effect of, or concomitant with, or all there is to my willing my hand to move, and which occurs before my hand moves. Even if one were skeptical of will being a cause, the brain activity would surely qualify and precede the effect.

For the second illustration a similar brain activity or act of will would qualify as the cause. And for the third illustration the cause is the other person moving the hand.

In the third case moving a hand and moving the pencil are both effects of a common cause. Are the other two cases examples of common cause, too? That question reduces to whether the inference from "I will my hand to move while grasping the pencil" to "The pencil moves" is valid or strong without the premise "The hand moves." But given the similarity of these cases to those in which we invoke a common cause, I suggest we extend the definition of *common cause* to include those cases where A may be required for the inference from C to B, when A and B are simultaneous. This preserves the temporal distinction between cause and effect.

*Example 37* In the case of boiling water, the boiling takes place just when the water reaches $100^\circ C$.; and so cause and effect are simultaneous.[43]

   *Analysis* By the extended definition given in the last example, the application of heat to the water is the common cause.

---

[43] James Ellington, Introduction to Kant's *Prolegomena*, p. xii. He also offers "Writing 'Larry' caused me to write the letter 'r' twice in succession," which is definition.

Richard Taylor gives as an example of simultaneous cause and effect the wind blowing causing a leaf on a tree to move.[44] He says that since time cannot distinguish cause from effect, some power or efficacy to produce changes in other things is what distinguishes cause from effect. This conception of power is supposed to be a metaphysical notion that cannot be analyzed in terms of necessary and sufficient conditions. He concludes:

> To say of anything, then, that it was the cause of something else, means simply and solely that it was the cause of the thing in question, and there is absolutely no other conceptually clearer way of putting the matter except by the introduction of mere synonyms for causation.[45]

This ties the notion of power to a metaphysics of things, which would not account for causal claims in which a substance such as mud, or a state of affairs, such as a drought, or a process, such as breathing, is said to be the cause.[46]

G. H. von Wright avoids "power" in favor of "manipulability":

> We now have a clue to answering what I called the *asymmetry problem*. This was the question of what distinguishes cause from effect . . . What makes $p$ a cause-factor relative to the effect-factor $q$ is, I shall maintain, the fact that by *manipulating p*, i.e. by producing changes in it "at will" as we say, we could bring about changes in $q$. . . .
>
> In the normal cases, the effect brought about by the operation of the cause occurs later. In such cases time has already provided the distinction. More problematic is the case when cause and effect are supposed to be simultaneous. Those who think of the cause-effect distinction in terms of temporality alone will be at a loss here. But when the distinction is made in terms of manipulability, the difficulty can be solved.[47]

---

[44] "The metaphysics of causation."
[45] Ibid., p. 43.
[46] Compare P. F. Strawson, *The Bounds of Sense*, pp. 145–146:

> [Our concepts of objects are] necessarily compendia of causal law or law-likeness, carry implications of causal power or dependence. Powers, as Locke remarked . . . make up a great part of our idea of substances. More generally, they make up a great part of our concepts of any persisting and re-identifiable objective items. And without some such concepts as these, no experience of an objective world is possible.

[47] "On the logic and epistemology of the causal relation," p. 118. See also C. J. Ducasse, "On the nature and observability of the causal relation."

But manipulability or "change" can only work by ignoring "static conditions" as causes: the drought caused the crops to fail. Even for an active condition such as a lightning it would be odd to say it is manipulable.

Bertrand Russell in "On the notion of cause" also challenges the view that cause must precede effect, not on the basis of cause and effect sometimes being simultaneous, but because time could run either way and the past is somehow equivalent to the future for causation.

To allow that cause and effect can be simultaneous creates serious problems for distinguishing cause from effect. But isn't the appeal to a common cause just a way to preserve the time distinction between cause and effect by fiat? After all, we can always trace the cause back farther. Since all the examples I've seen of simultaneous cause and effect can be dismissed by pointing to such obvious common causes or noting that it is a definition, I suggest that we preserve the temporal distinction.

*Example 38* Two electrons are beamed toward each other with the same large momentum. They interact with each other for a short time when they pass at known positions. Later, when we can be sure that the two electrons are so far apart they no longer have any interaction, we observe one of them, measuring its momentum. Then we can deduce the momentum of the other particle; but it is impossible to measure the position of the other particle. So measuring the momentum of the one electron caused it to be impossible to measure the position of the other particle, and this is simultaneous cause and effect.

*Analysis* Current physical theories tell us that the example can occur, and experiments have shown that it does seem to occur.[48]

The description of the cause cannot be "The momentum of the first electron was measured" because that by itself is not sufficient to give us a valid or strong inference to "The position of the second particle cannot be measured." Also needed is a description of the interaction of the particles and how they have had no further interaction either with each other or anything else, all of which clearly precede the effect and are hardly normal conditions.

The example was originally devised as a worry about the non-locality of cause and effect: the particles are too far apart to interact in

---

[48] See, for example, N. P. Landsman, "When champions meet: Rethinking the Bohr-Einstein debate."

the sense that no signal could go between them even at the speed of light in the time that we deduce that we cannot know the position of the second particle, yet our measurement in some sense amounts to an interaction between them.[49] Whether or how this might matter for requiring further conditions on good causal inferences under the rubric "The cause must be close in time and space to the effect" is an open question.

## Claims about the nature of cause and effect

*Example 39* Wanda getting really fat is a consequence of her huge appetite.

*Analysis* This is somewhat vague, but we can understand it as a causal claim with purported cause "Wanda is lot more hungry a lot more often than most people" and purported effect "Wanda is fat." With appropriate normal conditions, this might seem a plausible causal claim, though any doctor would look for a cause in her biology.

Hart and Honoré, however, suggest that terms such as "is a consequence" should not be so easily assimilated to cause and effect:

> Preoccupation with the familiar pair of terms "cause and effect" may make us think that there is a *single* concept of "causation" awaiting our inspection and that the huge range of other causal expressions, "consequence", "result", "caused by", "due to", "lead to", "made", etc., are mere stylistic variants. Sometimes indeed expressions of this group may be substituted without alteration of meaning. It matters little, for example, whether we say that a fire was *caused by* a short circuit, or was *due to* it or that a short circuit *led* to the fire. Yet very often this substitution cannot be made without change of meaning or gross incongruity of expression. The use of the term "effect" is in fact fairly definitely confined to cases where the antecedent is literally a *change* or *activity* of some sort (as distinct from a persistent or negative condition), and where the event spoken of as the effect is a change brought about in a person or continuing thing. Thus, though the icy condition of the road, or a failure to signal, may be the cause of an accident, this is not spoken of as the "effect" of these causes. Again, though we say of a person fined for driving a car at an excessive speed that this was the *consequence* of his speeding it is not the *effect* of it. The expression "the effect of speeding" calls to mind (because its

---

[49] As propounded by Albert Einstein; see L. Rosenfeld, "Niels Bohr in the thirties", pp. 127–128, and A. Einstein, B. Podolsky, and N. Rosen, "Can quantum-mechanical description of physical reality be considered complete?"

standard use is to refer to them) such things as the heating of the
engine, or the nervous fatigue of the driver. "Effects" as distinct from
"consequences" are usually "on" something, brought about as the
terminus of a series of changes which may or may not be deliberately
initiated by human agents but involve no deliberate intervention by
others. "Consequences" has a much wider application, though it
cannot always be used of the terminus of a very short series of
physical or physico-chemical changes. Hence the incongruity in
speaking of, for example, the melting point of wax in a flame as the
consequence of heating it: it is its effect. "Results" has also special
implications and so a characteristic sphere of application different
from either of the other two expressions. It is typically used of what
emerges as the culminating phase or outcome of a process which is
complex and consciously designed. So we speak of the "result" of a
game, a trial, an experiment. Here no substitution of "effect" or
"consequence" could be made without change of meaning. The
prisoner's acquittal is the *result* of a trial, its *effect* on the public may
be one of astonishment and the eventual change in the law may be one
of its *consequences*. It is clear that "cause and effect" have caught
too much of the limelight of philosophical attention; for they are in
fact not as frequently used as this has led us to think.

These few examples of what can be done to draw out differences
between various causal expressions perhaps suffice to show that there
is not a single concept of causation but a group or family of concepts.
These are united not by a set of common features but by points of
resemblance, some of them tenuous.[50]

These differences in linguistic usage do not signal different notions of
causation "united not by a set of common features but by points of
resemblance, some of them tenuous." All the examples can be treated
in the same framework of analysis of cause and effect in terms of inferences satisfying certain necessary conditions. Requiring further conditions on such inferences to justify the label "consequence" or "result"
could model Hart and Honoré's discussion.

*Example 40* Doctor: Your prostate gland is causing you to have
trouble urinating. You'll need surgery to remove it.
 *Analysis* This is how we talk. The prostate gland is the cause of
the trouble, and the physician will remove the cause. But if causes are
described by claims, how can you remove a cause?

---

[50] *Causation in the Law*, pp. 27–28; the discussion is continued on
pp. 55–56 there.

There are problems we get into when we talk of things or events as causes: metaphysical confusions and, above all, imprecision and confusion in reasoning about cause and effect. Here the cause is not a prostate gland, nor even a swollen or infected or cancerous prostate gland. We're not sure exactly what the cause is, because the physician hasn't said, but it's something like "Your prostate gland is swollen and is blocking your urinary tract," and the effect is "You can't urinate normally." To say colloquially that the physician will remove the cause is to say that she will remove the prostate gland which will make that first claim no longer true.

We do not need to respect the way we commonly speak if that way leads to or is a source of misunderstandings and bad reasoning. We can, however, continue to speak colloquially if we remember that more precision and care are needed when we want to reason about cause and effect.

*Example 41* If that piece of butter had been heated to $8°C$ for two hours, it would have melted.

*Analysis* This is a counterfactual causal claim, equivalent to:

Heating the piece of butter to $8°C$ for two hours, would have caused the butter to melt.

The conditions for it to be true cannot include that both the claim describing the cause and the claim describing the effect are true, for the point is that they aren't.[51] We can understand the claim as asserting that if the antecedent were true, then the consequent would follow by a good causal inference.

---

**Hypothetical causal claims** Suppose we have a causal claim about what would have happened in the past or might happen in the future where the purported hypothetical cause can be described by A and the purported effect by B. The necessary conditions for the causal inference to be good are that for every way in which A could have been or could become true:

- B becomes true at a later time than A.
- The inference from A to B is valid or strong.
- The inference from B to A is valid or strong.

---

[51] See the essay "Conditionals" in this volume for an analysis of counterfactual claims in which to place this discussion.

- There is no D that would describe a common cause of A and B which would be true in the hypothesized description of the world.

In practice, we take these necessary conditions to be sufficient.

*Example 42* Nothing causes itself.
*Analysis* How are we to understand this?
I suspect that what's intended is that nothing causes itself to exist. We can try to reason about this example with the methods of causal reasoning we've established. In that case it might seem that for the claim to be false there must be some object, say $a$, and some true description of the world in which $a$ does not exist, say A, such that "A, therefore $a$ exists" is a good causal inference. But that's not right, since it's a commonplace that such inferences exist, for example, when a puppy is born or an electron is created.

Rather, what appears to be meant by the example is that there is no description of the world W that somehow refers to $a$ in the present tense yet in which $a$ does not exist, such that "$W(a)$, therefore $a$ exists" is true. To analyze inferences of that sort we need not only rules for reasoning with non-referring names, but also how to reason with the predicate "exists." We need new rules of reasoning to deal with this example, if we can make sense of it at all.[52]

*Example 43* Everything has a cause.
*Analysis* How are we to understand this? From what I can gather from those who make this assertion, "everything" is meant to cover not only objects but any "event" or "action" or "state of the world." Analyzing causal claims in terms of causal inferences, this would mean:

For any claim B true of a certain time, there is some claim A true of an earlier time such that "A therefore B" is a good causal inference.

This no longer seems extraordinary. But to show that it is true we

---

[52] See my *The Internal Structure of Predicates and Names* for how to reason with non-referring names and the predicate "exists" in predicate logic.

We could make the example non-trivial and use our usual methods of reasoning by allowing cause and effect to be simultaneous. But that would be to make the example interesting only by using an assumption unneeded in other reasoning about cause and effect, as discussed in the previous section.

## 66  Cause and Effect

would need clearer criteria for what claims qualify for the roles that A and B play, and we would need sufficient criteria for an inference to be a good causal inference (a causal claim to be true), where we now have only necessary criteria.

Here is what Richard Taylor says of the usual discussions of this example:

> To assert that causation is universal is to assert that no change ever occurs without some cause—in short, that every event has a cause. . . . The universality of causation has throughout the history of philosophy, until very recent times, usually been regarded as very obvious, sometimes even self-evident. There are many thinkers today, however, who consider it quite possible that certain changes involving the minutest constituents of matter simply have no causes at all. . . . It is not the least difficult to imagine a change occurring without anything causing that change, and there is no contradiction whatever in asserting that this sometimes happens, although, of course, it may be false.[53]

Leibniz takes this example as one reading of his Principle of Sufficient Reason:

> The principle that a reason must be given is this: that every true proposition not known per se has an a priori proof, or that a reason can be given for every truth, or, as is commonly said, everything has a cause.[54]

The last phrase in the Latin original is *nihil fit sine causa*. Yet Leibniz usually says *nihil fit sine ratione*, "nothing comes to pass without a reason," where "reason" means something like the grounds for the truth of a claim (see Appendix 1 below).

Mill, too, discusses this example:

> The Law of Causation . . . is but the familiar truth, that invariability of succession is found by observation to obtain between every fact in nature and some other fact which has preceded it; . . . . To certain facts, certain facts always do, and, as we believe, will continue to succeed. The invariable antecedent is termed the cause; the invariable consequent, the effect. And the universality of the law of causation consists in this, that every consequent is connected in this manner with some particular antecedent or set of antecedents.[55]

---

[53] "Causation," pp. 57–58.
[54] In *Die Philosophischen Schriften*, ed. C. I. Gerhardt, Berlin, 1857–1890, vol. VII, 309. Translation by Benson Mates, *The Philosophy of Leibniz*, p. 155.
[55] *A System of Logic*, Book III, Chapter V, section 2 (pp. 236–237). He

Michael Scriven, however, denies one form of the law of causation:

> To say that "same cause, same effect" is a determinist's slogan is not to say it has *no* empirical content. It has, and it is actually false, as far as present evidence goes, though only to a small extent for macroscopic observations. ... What it lacks is single-case applicability and hence direct confirmability when complex systems are involved —for it is often impossible to specify what counts as the "same conditions".[56]

If the example has any meaning at all, it is a determinist's dream that has no obvious significance to reasoning about cause and effect beyond goading us to look for causes, which we would do anyway.

## Finding a cause

*Example 44* Evaporation and spray from my waterfall is causing the loss of water in my pond.

*Analysis* I have a waterfall in my backyard in Cedar City. The pond has a thick rubberized plastic pond liner, and I have a pump and hose that carry water from the pond along the rock face of a small rise to where the water spills out and runs down more rocks with concrete between them. Last summer I noticed that the pond kept getting low every day and had to be refilled. You don't waste water in the desert, so I figured I'd better find out what was causing the loss of water.

I thought of all the ways the pond could be leaking: The hose that carries the water could have a leak, the valve connections could be

---

gives four other formulations there, none of which appears to be equivalent:

> This law is the Law of Causation. The truth that every fact that has a beginning has a cause, is co-extensive with human experience.
>
> This generalization may appear to some minds not to amount to much, since after all it asserts only this: "it is a law, that every event depends on some law:" "it is a law that there is a law for everything." (Section 1, pp. 235–236)
>
> The Law of Causation, the recognition of which is the main pillar of inductive science, is but the familiar truth, that invariability of succession is found by observation to obtain between every fact in nature and some other fact which has preceded it, independently of all considerations respecting the ultimate mode of production of phenomena and of every other question regarding the nature of "things in themselves." (Section 2, p. 236)

[56] "Explanations, predictions, and laws," p. 189.

leaking, the pond liner could be ripped (the dogs get into the pond to cool off in the summer), there could be cracks in the concrete, or it could be evaporation and spray from where the water comes out at the top of the fountain.

I had to figure out which (if any) of these was the problem. First I got someone to come in and use a high pressure spray on the waterfall to clean it. We took the rocks out and vacuumed out the pond. Then we patched every possible spot on the pond liner where there might be a leak.

Then we patched all the concrete on the waterfall part and water-sealed it. We checked the valve connections and tightened them. They didn't leak. And we knew that the hose wasn't leaking because there weren't any wet spots along its path.

Then I refilled the pond. It kept losing water at about the same rate.

It wasn't the hose, it wasn't the connections, it wasn't the pond liner, it wasn't the concrete watercourse. So it had to be the spray and evaporation.

I reduced the flow of water so there wouldn't be so much spray. There was a lot less water loss. The rest I figured was probably evaporation, though there might still be small leaks.

In trying to find the cause of the water leak at my waterfall and pond I was using a method scientists often use.

*Finding a cause* Conjecture possible causes. By experiment eliminate them by showing that they don't make a difference, until there is only one. Check that one: Does it make a difference? If the purported cause is eliminated, is there still the effect? Is there a common cause?

I assumed there was a cause, then by a process of elimination on some conjectured causes, I fixed on one. When that occurred, the suspected effect always did, too, and it made a difference, and I knew I could fill in the normal conditions.

But why should I have assumed that there was a cause? Does this mean that I was assuming that everything has a cause? No, I was assuming that there is some way to stop the leak, which in this case amounts to assuming that the leak had a cause. The assumption that a particular effect has a cause is sometimes just an expression of our desire to find a way to manipulate the world.

But doesn't this method rest on a false dilemma?

A or B or C is the cause of E. It's not A. It's not B. Therefore, it's C.

No. We still have to check that C satisfies all the conditions for cause and effect, not just that it makes a difference. We must be willing to accept that our experiments might show that none of the conjectured causes satisfies all the conditions. This method cannot find the cause from nothing, but only, if we guess right, isolate it from a range of conjectured causes.

John Stuart Mill understands cause and effect in terms of invariable relations between things/events.[57] He formulated a method in five parts that he believed would serve both to discover invariable relations and prove that our discoveries were correct. Collectively they are called *Mill's methods of induction*:

*Method of Agreement* If two or more instances of the phenomenon under investigation have only one circumstance in common, the circumstance in which alone all the instances agree is the cause (or effect) of the given phenomenon.

*Method of Difference* If an instance in which the phenomenon under investigation occurs, and an instance in which it does not occur, have every circumstance in common save one, that one occurring in the former; the circumstance in which alone the two instances differ, is the effect, or the cause, or an indispensable part of the cause, of the phenomenon.

*Joint Method of Agreement and Difference* If two or more instances in which the phenomenon occurs have only one circumstance in common, while two or more instances in which it does not occur have nothing in common save the absence of that circumstance; the circumstance in which alone the two sets of instances differ, is the effect, or the cause, or an indispensable part of the cause of the phenomenon.

*Method of Residues* Subduct from any phenomenon such part as is known by previous inductions to be the effect of certain antecedents, and the residue of the phenomenon is the effect of the remaining antecedents.

*Method of Concomitant Variation* Whatever phenomenon varies in any manner whenever another phenomenon varies in some particular manner, is either a cause or an effect of that phenomenon, or is connected with it through some fact of causation.[58]

---

[57] See the quotes in Example 26 and footnote 26 above.

But these, too, do not help us discover a cause from nothing; they only isolate it from a range of conjectured possible causes. Mill's methods often seem more natural in terms of cause in populations, which are discussed in the next section.[59]

*Example 45* Recently Lee found out that he has hepatitis B. None of his friends has hepatitis. He wonders how he could have gotten it.

He reasons: Since he wants to be a nurse he volunteers to work at a hospital three times per week. Some of the patients there have hepatitis, and he often washes their bedpans and comes in contact with their body fluids, though he's always careful to wear gloves. Or at least he thought he was. A recent study he read said that 25% of health care workers exposed to hepatitis B get it. So, he figures, he got hepatitis B from working at the hospital.

*Analysis* How strong is this argument? We start with the conjecture: "Lee contracted hepatitis B from working at the hospital." We rule out all other causes we can think of. We can imagine conditions under which he could have gotten hepatitis, but we can't specify the exact conditions that occurred that would give us the normal conditions. Eliminating all other possible causes (that we can think of) doesn't mean that we can conclude we've found the cause unless we also have:

(*) The only ways Lee could have gotten hepatitis B are P, Q, R, S, T, U, or V.

There are very strong arguments that he didn't get it from Q, R, S, T, U, or V. Therefore, he got it from P. This reasoning to a cause is just as strong as (*) is plausible.

## Cause in populations

When we say smoking causes lung cancer, what do we mean? If you smoke a cigarette you'll get cancer? If you smoke a lot of cigarettes

---

[58] From *A System of Logic*, in order: Book III, Chapter VIII, section 1 (p. 280), section 2 (p. 280), section 4 (p. 284), section 5 (p. 285), and section 6 (p. 287).

[59] For an early critique of Mill's methods, see Chapter XIII of Morris Cohen and Ernest Nagel, *An Introduction to Logic and Scientific Method*. J. L. Mackie in *The Cement of the Universe* has both criticized these methods and shown how they can be sharpened to be of greater value; see Mackie's "Mill's methods of induction" for a shorter exposition and critique.

this week, you'll get cancer? If you smoke 20 cigarettes a day for 40 years you'll get cancer? It can't be any of these, since there are lots of people who smoke who did all that yet didn't get lung cancer, and the effect has to (almost) invariably follow the cause.

*Cause in a population* is usually explained as meaning that given the cause, there's a higher probability that the effect will be true than if the cause has not occurred. In this example, people who smoke have a much higher probability of getting lung cancer. But really we're talking about cause and effect just as we did before. Smoking lots of cigarettes over a long period of time will cause (inevitably) lung cancer. The problem is that we can't state, we have no idea how to state, nor is it likely that we'll ever be able to state the normal conditions for smoking to cause cancer. Among other factors, there is diet, where one lives, exposure to pollution and other carcinogens, and one's genetic inheritance. But *if we knew exactly* we'd say: "Under the conditions ___ , smoking ___ (number of ) cigarettes every day for ___ years will result in lung cancer."

Since we can't specify the normal conditions, the best we can do is point to the evidence that convinces us that smoking is a cause of lung cancer and get an argument with a statistical conclusion: "People who continue to smoke two packs of cigarettes per day for ten years are __% more likely (with margin of error __%) to get lung cancer."

Suppose we establish that smoking causes cancer in this statistical sense. That is not a general causal claim: there can be instances of someone smoking and not getting cancer, yet the general claim need not be abandoned on that account. Rather, the statistical claim serves as a premise: from this claim, and this person getting lung cancer after having smoked for thirty years, we conclude that the person's smoking was a cause of his getting cancer. *Cause-in-population claims help to justify particular causal claims, they do not summarize them.*

Perhaps, then, there are no general causal claims but only cause-in-population claims? No, the laws of Newton's mechanics for billiard balls are meant and seem to be true as general causal claims: Every ball hit by another under normal conditions will obey those laws. The particular causal claim which is an instance of the law will be true, and we can (more or less) specify those conditions.

There are three methods for establishing cause in a population.

*Controlled experiment: cause-to-effect*
This is our best evidence. We choose 10,000 people at random and ask 5,000 of them never to smoke and 5,000 of them to smoke a pack of cigarettes every day. We have two samples, one composed of those who are administered the cause, and one of those who are not, the latter called the *control group*. We come back 20 years later to check how many in each group got lung cancer. If a lot more of the smokers got lung cancer, and the groups were representative of the population as a whole, and we can see no other common thread among those who got lung cancer, we'd be justified in saying that smoking causes lung cancer.

Of course such an experiment would be unethical, so we use rats instead and then argue by analogy. Whether that is more ethical is another issue.

*Uncontrolled experiment: cause-to-effect*  Here we take two randomly chosen samples of the general population for which we have factored out other known possible causes of lung cancer, such as working in coal mines. One of the groups is composed of people who say they never smoke. The other group is composed of people who say they smoke. We follow the groups and 20 years later check whether those who smoked got lung cancer more often. Since we think we've accounted for other common threads, smoking is the remaining common thread that may account for why the second group got cancer more often.

This is a cause-to-effect experiment, since we start with the suspected cause and later see if the effect followed. But it is uncontrolled: Some people may stop smoking, some may begin, people have quite variable diets—there may be a lot we'll have to factor out in trying to assess whether it's smoking that causes the extra cases of lung cancer.

*Uncontrolled experiment: effect-to-cause*  Here we look at as many people as possible who have lung cancer to see if there is some common thread that occurs in (almost all) their lives. We factor out those who worked in coal mines, we factor out those who lived in high pollution areas, those who have cats, . . . . If it turns out that a much higher proportion of the remaining people smoked than in the general population, we have good evidence that smoking was the cause (the evaluation of this requires a knowledge of statistics). This is uncontrolled because how they got to the effect was unplanned, not within our control. And it is an effect-to-cause experiment because we start with the effect in the population and try to account for how it got there.

*Example 46*  Barbara smoked two packs of cigarettes a day for thirty years. Barbara now has lung cancer. Barbara's smoking caused her lung cancer.

*Analysis*  Is it possible for Barbara to have smoked two packs of cigarettes each day for thirty years and not get lung cancer? We can't state the normal conditions. So we invoke the statistical relation between smoking and lung cancer to say it is unlikely for the cause to be true and effect false.

Does the cause make a difference? Could Barbara have gotten lung cancer even if she had not smoked? Suppose we know that she wasn't a coal miner, didn't work in a textile factory, didn't live in a city with a very polluted atmosphere, and didn't live with a heavy smoker, all conditions that are known to be associated with a higher probability of getting lung cancer. Then it is possible for Barbara to have gotten lung cancer anyway, since some people who have no other risks do get lung cancer. But it is very unlikely, since very few of those people do.

We have no reason to believe that there is a common cause. Maybe people with a certain biological make-up feel compelled to smoke, and that biological make-up also contributes to their getting lung cancer independently of their smoking. But we've no evidence for that, and before cigarette smoking became popular lung cancer was rare.

So assuming a few normal conditions, "Barbara's smoking caused her lung cancer" is as plausible as the strength of the statistical link between smoking and lung cancer, and the strength of the link between not smoking and not getting lung cancer. We must be careful, though, that we do not attribute the cause of the lung cancer to smoking just because we haven't thought of any other cause, especially if the statistical link isn't very strong.

*Example 47*  Zoe: I can't understand Melinda. She's pregnant and she's drinking.
Dick: That's all baloney. I asked my mom, and she said she drank when she was pregnant with me. And I turned out fine.
Zoe: But think how much better you'd have been if she hadn't.

*Analysis*  Zoe doesn't say but alludes to the cause-in-population claim that drinking during pregnancy causes birth defects or poor development of the child. That has been demonstrated: Many cause-in-population studies have been done that show there is a higher incidence of birth defects and developmental problems in children born to women who drink during pregnancy than to women who do not drink, and those defects and problems do not appear to arise from any other common factor.

Dick, however, makes a mistake. He confuses a cause-in-population claim with a general causal claim. He's right that his mother's experience would disprove the general causal claim, but it has no force against the cause-in-population claim.

Zoe's confusion is that she thinks there is a perfect correlation between drinking and physical or mental problems in the child, so that if Dick's mother had not drunk he would have been better, even if Zoe can't point to the particular way in which Dick would have been better. But the correlation isn't perfect; it's only a statistical link.

*Example 48* Lack of education causes poverty. Widespread poverty causes crime. So lack of education causes crime.

*Analysis* We often hear words like these, and some politicians base policy on them. But they're too vague. How much education constitutes "lack of education"? How poor do you have to be? How many poor people constitute "widespread poverty"? Researchers make these sentences more precise and analyze them as cause-in-population claims, since we know they couldn't be true general causal claims: There are people with little education who've become rich; and lots of poor people are law-abiding citizens. Indeed, during the worst years of the Depression in the 1930s, when there was more widespread poverty than at any time since in the U.S., there was less crime than any time in the last 20 years. This suggests it would be hard to find a precise version of the second sentence that is a true cause-in-population claim.

*Example 49* Several studies indicate that people who smoke cigarettes
have an increased risk for low back pain and prolapsed disk [references given]. Individuals who have not smoked for more than a year, however, do not appear to have an increased risk, as least for prolapsed lumbar disk [reference given]. Table 6 shows that current smokers have almost twice the risk for prolapsed lumbar disk as those who have never smoked or who are former smokers. In the same study [reference given] it was estimated that the risk in current smokers is increased by about 20% for every 10 cigarettes smoked per day on the average. Possible mechanisms for the association between smoking and low back

pain and prolapsed disk include decreased diffusion of nutrients into the intervertebral disk among smokers [reference given], and increased pressure on the low back from the frequent coughing experienced by many smokers.

Table 6. Estimated Relative Risk for Prolapsed Lumbar Intervertebral Disk According to Cigarette Smoking Status, Connecticut

| Smoking Status | Estimated Relative Risk* | 95% Confidence Limits |
|---|---|---|
| Never smoked (referent group) | 1.0 | — |
| Current smoker (smoked in past year) | 1.7 | 1.0–2.5 |
| Former smoker (smoked, but not in past year) | 1.0 | 0.6–1.7 |

*Relative risk = risk in those exposed to factor divided by risk in those not exposed (referent group).[60]

*Analysis*   The authors suggest that the cause-in-population studies they cite show smoking causes lower back pain. But perhaps they've got cause and effect reversed: people who have back pain might want to smoke to take their minds off the pain, or possibly even to alleviate the pain. Or there could be a common cause: people who do manual labor might smoke more, and the manual labor also causes back problems. Until further cause-in-population studies rule out those possibilities, this is just a *correlation-causation fallacy*: claiming that a correlation by itself establishes cause and effect; that's just a pumped-up version of *post hoc* reasoning or reversing cause and effect.

## Causal laws

In the analysis of cause and effect presented here, particular causal claims take priority in the sense that we do not need to establish the general in order to establish the particular. C. J. Ducasse argues similarly:

> The definition of cause proposed . . . defines the cause of a particular event in terms of but a single occurrence of it, and thus in no way involves the supposition that it, or one like it, ever has occurred before or ever will again. The supposition of recurrence is thus wholly irrelevant to the meaning of cause; that supposition is relevant only to the meaning of law. And recurrence becomes related at all to

---

[60] Jennifer L. Kelsy, Anne L. Golden, Diane J. Mundt, *Rheumatic Disease Clinics of America*, vol. 16, no. 3, 1990.

causation only when a law is considered which happens to be a generalization of facts themselves individually causal to begin with. ... The causal relation is essentially a relation between concrete individual events; and it is only so far as these events exhibit likeness to others, and can therefore be grouped with them into kinds, that it is possible to pass from individual causal facts to causal laws.[61]

This is rejected by many. Morris R. Cohen and Ernest Nagel say the general causal claim has priority:

> By the *cause* of some *effect* we shall understand, therefore, some appropriate factor invariably related to the effect. If *A has diphtheria at time t* is an effect, we shall understand by its cause a certain change *C,* such that the following holds. If *C* takes place, then *A* will have diphtheria at time *t,* and if *C* does not take place, *A* will not have diphtheria at time *t* ; and this is true for all values of *A, C,* and *t,* where *A* is an individual of a certain type, *C* an event of a certain type, and *t* the time.[62]

This is in the spirit of Hume: constant conjunction of "similar" causes to "similar" effects is necessary for us to conclude that there is cause and effect. But that requirement is too strong: many of the examples above were *sui generis*, yet we had no compunction in labeling them cause and effect.

Perhaps the priority of the general causal claim is just the requirement that some general principle must be invoked as a premise for a causal inference to be good. Even if we do not accept Hume's analysis, we can ask whether every good causal inference must have a universal claim as premise. Those who say "yes," the *regularists*, call such universal claims "causal laws." Hart and Honoré say:

> Hume's insistence that constant conjunction or regular sequence between events is the essence of the notion of causation is represented by the doctrine that every singular causal statement implies, by its very meaning, a general proposition asserting a universal connection between kinds of events; to make such a singular causal statement is therefore to claim that the events which it relates are instances of such a universal connection between types of events. The psychological analysis of the idea of necessary connection between cause and effect

---

[61] "On the nature and the observability of the causal relation," pp. 129–130.
[62] *An Introduction to Logic and Scientific Method*, p. 248. Here the effect is taken to be an open sentence which when instantiated becomes a proposition, though Cohen and Nagel then talk about causes as events or changes.

has undergone a more drastic change; it is generally accepted that, though Hume was right in insisting that necessity does not "lie in objects", not all events which follow each other in invariable sequence are causally related, and not every general proposition asserting such sequence between kinds of events is ranked as a "causal law". Those that are, are distinguished not by the "feeling" which they engender in the mind, but more by the place they occupy in a set of related general propositions, or within a scientific theory, or by the fact that they are used, as many other generalizations are not, to justify inferences, not merely as to what has happened or will happen, but "counterfactual" inferences as to what *would* have been the case if some actual event, which in fact happened, had not happened.[63]

What makes a claim a "causal law" as opposed to just a premise that strengthens a causal inference? Compare:

(a) Every coin I have in my pocket is U.S. currency.

(b) Every electron has negative charge.

The first is labeled a contingent universal claim by regularists, what is sometimes called an *accidental generalization*. It does not "support" counterfactuals, for (a) could be true while "If this coin were in my pocket it would be U.S. currency" could be false. On the other hand, regularists say that (b) is a causal law. It supports counterfactuals: if (b) is true, then so is, "If this were an electron, it would have negative charge."

The difference, then, seems to be that a contingent universal claim just "happens" to be true. It is true, but it says nothing about possibilities. A causal law, on the other hand, speaks of possibilities: it is not only true, but true for all possibilities; it is necessary. However, regularists typically qualify the kind of necessity to cover only "causally possible worlds."[64] But possibilities are already taken into

---

[63] *Causation in the Law*, p. 15.

[64] See Ernest Nagel, *The Structure of Science*, pp. 52–56 for a discussion of the kind of possibilities surveyed in causal laws. If in specifying what is a causal law we invoke physically possible worlds or causally possible worlds, then there is a risk of circularity, for laws are used to determine what is physically possible or causally possible. The former, however, could be explained by reference to particular physical laws (but which ones is the great difficulty), relying on our ability to spot a law when we see one, with no need for a general definition of "law." See also the discussion of causal laws by William A. Wallace in Chapter 4.3.b of *Causality and Scientific Explanation*.

consideration in that a causal inference must be valid or strong, and the inference from effect to cause must also be valid or strong for the inference to be good.

Morton White makes a criticism of the regularist's position.[65] It may be that the general claim which is needed is "All A are B" in order to get that *a* is B. But especially in historical studies, often the only instance of an A is *a*, so that the inference from "*a* is A" to "*a* is B" doesn't get better by invoking "All A are B." White says that this shows the inference is weak and doesn't get better by adding the universal claim; requiring the universal claim makes it clearer what's wrong with the original inference. That may be so for his historical examples, but the causal inferences in the examples we've seen weren't weak due to a lack of a universal claim as premise; they just weren't valid. Cause-in-population claims often serve as the general principle in causal inferences. That reflects our acceptance of strong causal inferences as good and shows that the "causal law" need not be universal. The issues are not significantly different from whether we allow good arguments to be only strong.

The regularist needs to be able to distinguish law-like claims from accidental generalizations, yet so far no generally accepted criteria have been proposed.[66] Perhaps Nicholas Rescher is right in explaining why it is so difficult to characterize what we mean by a "law" (of nature):

> Lawfulness in not found in or extracted from the evidence, it is super-added to it. *Lawfulness is a matter of imputation.* When an empirical generalization is designated as a law, this epistemological status is *imputed* to it. Lawfulness is something which a generalization could not *in principle* earn entirely on the basis of warrant by the empirical facts. Men impute lawfulness to certain generalizations by according to them a particular role in the epistemological scheme of things, being

---

[65] *Foundations of Historical Knowledge*, pp. 22ff.

[66] Certainly it's not that every law can be formalized as a universal or statistical claim in predicate logic or some simple extension of that logic, which is the assumption that infects much of the discussions in Carl G. Hempel, "Aspects of scientific explanations," Morton White, *Foundations of Historical Knowledge,* and Ernest Nagel, *The Structure of Science* (especially p. 32 and p. 47). A general principle might be "All gold is valuable" or "Water is $H_2O$" or "Almost all dogs bark," none of which can be formalized in predicate logic. See "The metaphysical basis of logic" in *Reasoning and Formal Logic* in this series.

prepared to use them in special ways in inferential contexts (particularly hypothetical contexts), and the like.

When one looks at the explicit formulation of the overt *content* of a law all one finds is a certain generalization. Its lawfulness is not a part of what the law asserts at all; it is nowhere to be seen in its overtly expressed content as a generalization. Lawfulness is not a matter of what the generalization *says*, but a matter of *how it is used*. By being prepared to put it to certain kinds of uses in modal and hypothetical contexts, it is *we*, the users, who accord to a generalization its lawful status thus endowing it with nomological necessity and hypothetical force. Lawfulness is thus not a matter of the assertive content of a generalization, but of its epistemic status, as determined by the ways in which it is deployed in its applications.

Hume maintained that faithfulness to the realities of human experience requires us to admit that we cannot find nomic necessity in nature. Kant replied that necessity does indeed not reside in observed nature but in the minds of man, which projects lawfulness into nature in consequence of features indigenous to the workings of the human intellect. Our view of the matter agrees with Hume's that lawfulness is not an observable characteristic of nature, and it agrees with Kant that it is a matter of man's projection. But we do not regard this projection as the result of the (in suitable circumstances) inevitable working of the epistemological faculty-structure of the human mind. Rather, we regard it as a matter of *warranted decision*, a deliberate man-made imputation effected in the setting of a particular conceptual scheme regarding the nature of explanatory understanding.[67]

Nonetheless, there may be reasons for our characterizing a claim as a law (God wills it, nature is that way, . . . ) even if we cannot determine them. We need not deny realism.

Norwood Russell Hanson says,

> One could predict my going to sleep from watching me wind the clock, or retrodict my having wound the clock from observing me asleep. But this is risky, like amateur weather-forecasting or angling advice. No conceptual issue is raised by the failure of such a prediction or retrodiction. Our understanding of nature receives no jar from guessing wrong here on one occasion.
>
> This shows what we expect of a causal law. . . . It is not merely that no exceptions have been found. We are to some extent conceptually unprepared for an exception: it would jar physics to its foundations; the

---

[67] *Scientific Explanation*, p. 107 and p. 113, italics in the original.

pattern of our concepts would warp or crumble. This is not to say that exceptions do not occur, but only that when they do our concepts do warp or crumble.[68]

Even if this criterion could be used to distinguish laws from accidental generalizations, it is based on a mistaken view of the role of laws in science: they are not true, but only aids in reasoning that are applicable and which define their subject matter.[69]

Whether a law is needed as premise is part of the general problem of whether for any inference that is not a generalization or does not follow by only form or the meaning of words to be good there must a general principle among its premises.[70] The main reason to make a "general law" explicit in a causal inference, at least at present, is the same one we always have in dealing with inferences: to make clearer that the inference is valid or strong.[71]

## Conclusion

The mystery of cause and effect can be circumvented if not eliminated in our reasoning by using claims to describe purported causes and purported effects and understanding a causal claim as true if and only if the relation between those claims satisfies the conditions for a good causal inference. Different notions of cause and effect correspond to placing different conditions on what counts as a good causal inference. This provides a method of reasoning about cause and effect that is clear and useful in both our ordinary lives and science.

---

[68] *Patterns of Discovery*, p. 64.

[69] See "Models and theories" in *Reasoning in Science and Mathematics* in this series.

[70] See "Arguments" in the *The Fundamentals of Argument Analysis* also in this series.

[71] See also the sections on causal explanations and laws in the essay "Explanations" in this volume.

## Appendix 1  Causes and reasons

Until we attached metaphysical assumptions about time and objects to the notion of cause, causes and reasons, in the sense of grounds for the truth of a claim, seemed to be in the same category. Though it may seem wrong now to conflate causes and reasons, many earlier philosophers did. Socrates says in Plato's *Phaedo*, 100c–d:

> I cannot understand these other ingenious theories of causation. If someone tells me that the reason why a given object is beautiful is that it has a gorgeous color or shape or any other such attribute, I disregard all these other explanations—I cling simply and straightforwardly and no doubt foolishly to the explanation that the one thing that makes the object beautiful is the presence in it or association with it, in whatever way the relation comes about, of absolute beauty.[72]

Aristotle spoke similarly in his discussion of the fallacy of false cause (see p. 90 below). He recognized four kinds of causes:

> Now that we have established these distinctions, we must proceed to consider causes, their character and number. Knowledge is the object of our inquiry, and men do not think they know a thing till they have grasped the "why" of it (which is to grasp its primary cause). ...
>
> In one sense, then, (1) that out of which a thing comes to be and which persists, is called "cause", e.g. the bronze of the statue, the silver of the bowl, and the genera of which the bronze and the silver are species.
>
> In another sense (2) the form or the archetype, i.e. the statement of the essence, and its genera, are called "causes" (e.g. of the octave the relation of 2:1, and generally number), and the parts in the definition.
>
> Again (3) the primary source of the change or coming to rest; e.g. the man who gave advice is a cause, the father is cause of the child, and generally what makes of what is made and what causes change of what is changed.
>
> Again (4) in the sense of end or "that for the sake of which" a thing is done, e.g. health is the cause of walking about. ("Why is he walking about?" we say. "To be healthy", and having said that, we think we have assigned the cause.) The same is true also of all the intermediate steps which are brought about through the action of something else as means towards the end, e.g. reduction of flesh, purging, drugs, or surgical instruments are means towards health. All these things are "for the sake of" the end, though they differ from one another in that some are activities, others instruments.

---

[72] See also the quotation from Plato's *Phaedo* in Example 25.

This then perhaps exhausts the number of ways in which the term "cause" is used. . . . All the causes now mentioned fall into four familiar divisions. The letters are causes of syllables, the material of artificial products, fire, &c., of bodies, the parts of the whole, and the premises of the conclusion, in the sense of "that from which".[73]

These are now called (1) the "material cause," (2) the "formal cause," (3) the "efficient cause," and (4) the "final cause." Gregory Vlastos says that it is inappropriate to translate the Greek word that Aristotle used, αιτια ("aitia"), as "cause."

> To say that X is the αιτια of Y comes to precisely the same thing as saying that Y happened, or happens, or is the case, *because* of X. In proof of this, if proof it needs, I need only refer to the fact that Aristotle refers to his *aitiai* as "all the ways of stating το δια τι (the *because*)": Aristotle's four "causes" are his four "*be*causes."[74]

Vlastos goes on to show that many of the uses of *aitia* in Aristotle would not be causes as we understand them. We would be better, he says, to translate *aitia* as "reason" or "that on account of which", rather than thinking that Aristotle was so clearly mistaken about the nature of cause and effect. Vlastos was convinced that there was no way that Aristotle's discussion of *aitia* could be fully compatible with our notion of cause and effect.[75]

Leibniz, too, seemed to conflate what we understand by cause and reason in his discussion of "principle of causation":

> All truths—even the most contingent—have an a priori proof, or some reason why they are truths rather than not. And this is just what is meant when it is commonly said that nothing happens without a cause, or, that there is nothing without a reason.[76]

Yet Leibniz does say that not every reason is a cause; causes, it seems, are a subclass of reasons.

*Axioma Magna*
Nothing is without its reason.
   Or, what is the same, nothing exists without its being possible (at least for an omniscient being) to give a reason why it exists rather than not, and why it is in this condition rather than some other.

---

[73] Aristotle, *Physica,* Book II, Chapter 3 (194:15–195:19).
[74] "Reasons and causes in the *Phaedo,*" p. 79, italics in the original.
[75] Julius Moravcsik in "Aristotle on adequate explanations" argues that we should view Aristotle's four kinds of causes as four kinds of explanations.
[76] *Die Philosophische Schriften von. G.W. Leibniz,* VIII, p. 301, as translated by Parkinson.

*Example B* Zoe jogs every day because of her health.
   *Analysis* An Aristotelian would say that the health of Zoe is the final cause of her jogging. But we do not need any new metaphysical assumptions here. Rather, we can say that the cause is "Zoe wants to be healthy" and the effect is "Zoe jogs every day."

*Example C* Birds have wings so that they can fly.
   *Analysis* An Aristotelian would say that birds flying is the final cause of their having wings. This would be a case of the cause occurring after the effect. But evolutionists have shown that no appeal to final causes or a direction in which the universe is headed is needed here, only natural selection.

*Example D* Man is a rational being so that he can appreciate the greatness of God.
   *Analysis* This is a general causal claim. A particular instance of it would invoke an Aristotelian final cause "Meher Baba can appreciate the greatness of God" with effect "Meher Baba is a rational being." This is a final cause that apparently cannot be reduced to the metaphysics used for the necessary conditions of the usual notion of cause and effect. The cause becomes true after the effect.

   An Aristotelian might say that Examples B and C were true examples of teleological causation, too: effects determined by what goal will be accomplished in the future. To disagree is to adopt a different metaphysics, but it is to adopt a different metaphysics within the framework we have established with the minimal notion of cause and effect.

## Appendix 2  Events

Often causes and effects are said to be states of affairs, or situations, or events. But what is the state of affairs that makes "Juney was barking" true? Simply, Juney was barking. What situation is described by "Ralph is a dog"? Simply, Ralph is a dog. What event is described or makes "Dick woke up" true? Simply, Dick woke up.

   Claims are what we use to describe "the world," however you construe that. We have claims and we have the world. If we have in addition events, we need to distinguish one event from another. But that is very hard without resorting to claims that describe the events.[81]

---

[81] Nicholas Unwin in "The individuation of events" presents a survey of this problem. Donald Davidson in "Causal relations" says events are needed to clarify and to give the truth conditions of causal claims, since we apparently talk of events in our ordinary speech. He says that we need events as things, because otherwise we wouldn't be able to give the logical forms of causal claims, meaning predicate logic forms. But after surveying all the possibilities

# 86   Cause and Effect

Richard Montague believed that events are things and that we must allow for inexpressible events, because claims such as the following are true:[82]

> The same thing happened today as yesterday.
> Not everything's happened yet.

It's the same, he says, as why we must allow for inexpressible propositions. It certainly is because events are described by (expressed with, picked out by) claims, just as abstract propositions (if you believe in them) are picked out or expressed or represented by claims. In the end Montague never says what an event is but rather:

> The event "corresponds" to . . . a piece of language.
> We might identify the event with . . . a piece of language.[83]

J. L. Mackie invokes not only events, but states and generic events:

> Mill includes in his assemblages of conditions *states*, that is, standing conditions, as well as what are strictly speaking *events*. He also stresses the importance of factors which we should naturally regard as negative, for example, the absence of a sentry from his post. . . . If a certain type of event is symbolized as $C$, then not-$C$, or $\overline{C}$, will be the absence of any event of that type. . . .
> What is typically called a cause . . . may be a state rather than an event.[84]

---

for criteria of individuating events, Davidson in "The individuation of events" comes to the conclusion that the best criterion we can muster is that events are different if and only if they differ in their causes and/or effects. That is, we need events to explain cause and effect, but we need causes and effects to understand what an event is.

[82] "On the nature of philosophical entities."
[83] Ibid, p. 149.
[84] *The Cement of the Universe*, pp. 62–64, where he also introduces facts:

> A man sets out on a trip across the desert. He has two enemies. One of them puts a deadly poison in his reserve can of drinking water. The other (not knowing this) makes a hole in the bottom of the can. The poisoned water all leaks out before the traveller needs to resort to this reserve can; the traveller dies of thirst. p. 44
>
> The matter can be clarified if we introduce here a distinction between *facts* and *events* both as causes and as results or effects. In [this example] the puncturing of the can brought it about that the traveller died of thirst, that is, it caused his dying of thirst (though it prevented his dying of poison). But we cannot say that the puncturing of the can brought it about that he died, or caused his dying—since he would have died anyway, if it had not been punctured. *That he died*, and

This becomes simpler as a classification of kinds of claims. It would be interesting to try to make precise what it means for a claim to express a static condition.

Jaegwon Kim's explanation of events is meant as an improvement on Mackie's:

> We need entities that possess both an element of generality and an element of particularity; the former is necessary for making sense of the relations of necessity and sufficiency, and the latter for making sense of singular causal judgements.
>
> Such entities are ready at hand, however, since realizations of properties at particular space-time regions or objects (if one accepts some sort of substance ontology) fill the bill; they are general in that they involve properties, and particular in that they involve particular space-time regions or objects. Thus, we take an event to be the exemplifying of an empirical property by an object at a time (alternatively, at a space-time region, but we shall adopt the former approach).[85]

Kim seems to be trying to make events correspond to what is described by atomic claims, based on an Aristotelian or predicate logic metaphysics of substances or objects. In any case, when events in Kim's sense are used in reasoning, we find it is either a claim or something very much like a claim that is used to stand in for the event.

David Lewis' view is similar to Kim's, but Lewis takes events to be abstract objects which nonetheless have a position and time:

> I propose to identify events with their corresponding properties. An event is a property, or in other words a class, of spatio-temporal regions.
>
> An event occurs in a particular spatiotemporal region. Its region might be small or large; there are collisions of point particles and there are condensations of galaxies, but even the latter occupy regions small by astronomical standards.[86]

> *that he died of thirst*, are distinguishable facts, and hence distinguishable results. So, as long as we are dealing with fact-results, it is not surprising that the puncturing of the can should have brought about the second of these but not the first. But if we think of an effect as a concrete event, then the event which was the traveller's death was also his death from thirst, and we must say that the puncturing of the can caused it, while the poisoning did not. p. 46

[85] "Causes and events: Mackie on causation," p. 71, which also contains a serious criticism of Mackie's talk of events and conditions playing logical roles in Mackie's analysis of causation.
[86] "Events," p. 245 and p. 243.

But what space-time region is to be associated with an event or state of affairs? As Benson Mates pointed out to me, "The cat is on the mat" is supposed to be made true by or describe a state of affairs or an event. But what is included in that event? The cat is touching the mat? The cat is upon the mat? The earth, because up and down can only be determined relative to that? Where do we stop? Probably only at the entire universe. But simply, the event is that the cat is on the mat. We use "that" to restate the claim.

We have good criteria for individuating claims: the same words in the same order with the same punctuation. We draw equivalences between claims. Some say we do so because different claims can "express the same proposition," as some would say we need events because all the following describe the same event:

Spot was barking.
Spot was barking at 3 a.m.
Spot was yapping loudly.
Spot barked at the moment he saw a raccoon.

All of these would qualify as "the cause" of Dick waking up. But there cannot be so many different causes. They're all the same. They're descriptions of the same event.[87]

Events are used to justify our drawing equivalences. But we can draw equivalences without events. We identify claims because we believe they have the same truth-conditions, though those truth-conditions may be informal or indeed unstated. And those truth-conditions do not need states of affairs or events or situations.[88]

Causes and effects, whatever they may be, can be described or identified with claims, and that is what we need in order to reason about cause and effect.

---

[87] Igal Kvart, in "Cause and some positive causal impact," takes events as fundamental:

> Events, as employed in this paper, are narrowly individuated.
> [footnote] E.g., along the lines of triples of an object, a predicate (or property) and a time, for a simple case; or alternatively, taken under descriptions. (p. 405, 438).

Each example he gives of an event is with a claim, e.g., "$x$'s finger was cut off at $t_1$ by a machine" (p. 406). Yet he was adamant in a lecture in February 2000 that (logically?) distinct claims always describe distinct events, so that each of these ways of describing what we see in the picture of Spot and Dick on the second page of this essay is a distinct event.

[88] See my *Predicate Logic* and *The Internal Structure of Predicates and Names* where the truth of atomic claims is taken as primitive. See also the essay "Truth" in *The Fundamentals of Argument Analysis* in this series.

Only if we take things *simpliciter* as causes, things simply existing, without place or time or action, might a cause not be describable by a claim.[89] But such causes also would not be events or situations.

## Appendix 3  Objective chance

Some authors ascribe a notion of objective chance to objects or events that could account not only for strong but weak inferences being causal.[90] Here is what J. L. Mackie says:

> Objective chance is a counterpart of the power or necessity in causes for which Hume looked in vain. Just as the power in a cause would be something present in every instance of a certain kind of cause which somehow *guaranteed* the subsequent occurrence of the corresponding effect, so a penny's having, at each toss, a certain chance of falling heads and a certain (perhaps different) chance of falling tails would be something present in the initial stages of every individual tossing process which *tended* to produce the result "heads" and *tended* to produce the result "tails", where these tendencies might be either equal or unequal. ... It may make things a bit clearer if we say that objective chance, in the primary and strongest sense, would be an indeterministic counterpart of causal necessity.[91]

The difficulty is how objective chance can be involved in our reasoning about cause and effect. Brian Skyrms says,

> Those who think of chance as an irreducible notion of physical tendency or propensity usually think of it as a theoretical concept on a par with the concept of force in physics. Physical theories stated in terms of chance permit predictions about the chances. But all that we can observe are the relative frequencies in sequences of outcomes. These are inductively rather than deductively linked to statements about chance. That is, our rational degrees of belief about the chances are influenced by observed relative frequencies. Conversely, our beliefs about the chances influence our anticipations about relative frequencies. In short, all we know about chance in general is its connection with rational degree of belief and relative frequency. The main shortcoming of this view is that it has so little to say.[92]

---

[89] Alfred North Whitehead, in *Science and the Modern World*, has a richer notion of event that does not seem amenable to being described propositionally. Unfortunately, he does not explain how one could use that notion in reasoning.
[90] See David Lewis, "A subjectivist's guide to objective chance" and Igal Kvart, *A Theory of Counterfactuals* and "Cause and some positive causal impact."
[91] *Truth, Probability and Paradox*, pp. 179–180.

## 90 Cause and Effect

Nonetheless, the role of objective chance can be modeled with the inferential approach presented here.[93]

## Appendix 4  Is it new to analyze the relation of cause to effect as an inference?

Aristotle seems to have had the view of a cause as a premise:

> The refutation which depends upon treating as cause what is not a cause, occurs whenever what is not a cause is inserted in the argument, as though the refutation depended upon it. This kind of thing happens in arguments that reason *ad impossibile*: for in these we are bound to demolish one of the premisses. If, then, the false cause be reckoned in among the questions that are necessary to establish the resulting impossibility, it will often be thought that the refutation depends upon it.[94]

This was part of his understanding of causes as certain types of reasons (see Appendix 1). What about today, when most reject that view?

Something like the analysis of the relation of cause to effect as an inference is implicit in the writing of many modern authors. It is often said that one event is the antecedent or consequent of another, or that one event is necessary, or a necessary condition, for another, though still considering causation to hold between objects or some notion of event. A. C. Ewing says:

> Such a connection [between cause and effect] I can only think at all definitely by thinking cause and effect as connected by a relation of logical entailment, i.e. as internally related in the tenth sense of this term.[1]
>
> [1] I. e. so related that one is logically dependent for its existence on the other, and that it would be logically impossible for the one to be what it is without the other.[95]

Brand Blanshard says that the relation of cause to effect is a necessary connection, but the necessity he is speaking of resides in things:

> Being causally connected involve[s] being connected by a relation of logical *necessity*. ... Let *a* and *x* be *any* two things in the universe. They are then related to each other causally. But if

---

[92] *Choice & Chance*, p. 212.

[93] See the appendix of "Conditionals" in this volume for a further discussion of objective chance. Probabilities are discussed more fully in "Probabilities" in *The Fundamentals of Argument Analysis* in this series.

[94] *De Sophisticis Elenchis*, 167b, 21–27.

[95] *Idealism*, p. 171.

causally, then also intrinsically, and if intrinsically then also
necessarily, in the sense that they causally act as they do in virtue of
their nature of character, and that to deny such activity would entail
denying them to be what they are.[96]

Max Born says that he does not understand causal dependence as logical dependence, but as "dependence of real things of nature on one another."[97] Then he says:

Causality postulates that there are laws by which the occurrence of an entity $B$ of a certain class depends on the occurrence of an entity $A$ of another class, where the word "entity" means any physical object, phenomenon, situation, or event. $A$ is called the cause, $B$ the event.[98]

Some sense must be made of what Born means by "object, phenomenon, situation, or event" and how those are to be treated as somehow on a par for this notion of dependence. To me the obvious way is to describe causes with claims and then the use of laws to derive one from another is much like what I propose.

Arthur Burks in *Chance, Cause, Reason* proposes understanding causal necessity as an extension of logical necessity. He defines:

$\Box^c A$, read "A is causally necessary," is true iff A is true in every causally possible universe. Here A is a closed formula in the language of first-order predicate logic.

$A \to_c B$, read as "A causally implies B," is equivalent to $\Box^c(A \supset B)$, which is true iff in every causally possible universe $A \supset B$ is true.

The causally possible universes are a subset of the logically possible universes (worlds). Thus, it seems, Burks takes first-order predicate logic formulas, or what they stand for, to be causes and effects. Since he does not allow for iterated modalities, his approach is just a formal version of the idea that for A to cause B, the inference from A to B must be valid relative to the laws that define what is causally possible. But a single premise inference is often not valid even in this notion of validity; normal conditions have to be assumed. So Burks adds further defined connectives, particularly "A ec B" to be read as "A elliptically causes B," which essentially says that there is some condition (premise) such that $\Box^c(A \supset B)$ is true. It is hard to evaluate Burks' ideas, though, because he gives no ordinary language examples. Certainly Burks' approach does not allow for discussion of the possibility that the relation of cause to effect may be only strong or for the possibiilty of a cause not being describable in predicate logic.

---

[96] *The Nature of Thought*, vol. II, pp. 515–516. Italics in the original.
[97] *Natural Philosophy of Cause and Chance*, p. 6.
[98] Ibid., p. 9.

Those who understand explanations as deductions must take the relation of cause to effect either to be or to parallel a kind of inference in order to deal with causal explanations. Karl R. Popper says:

> To give a *causal explanation* of an event means to deduce a statement which describes it, using as premises of the deduction one or more *universal laws*, together with certain singular statements, the *initial conditions*. ... The initial conditions describe what is usually called the *"cause"* of the event in question.[99]

Carl G. Hempel says that cause and effect is an inference relation, though his analysis requires that there be "general laws" and is intertwined with his requirement that statements be about "observables."[100]

Similarly, Morton White adopts a thesis that seems like my analysis:

> Thesis I: A statement of the form "*A* is a contributory cause of *C*" is true if and only if there is an explanatory deductive argument containing "*A*" as premise and "*C*" as its conclusion.[101]

But then White says:

> Singular causal statements are not arguments at all but statements.
>
> A singular explanatory statement of the kind analyzed in Thesis I is true if and only if a certain kind of explanatory argument exists, but the argument is distinct from the statement, and the statement is not elliptical for the argument.[102]

I agree that a causal claim is not the same as a causal inference: a causal claim can be analyzed as an assertion that an inference satisfies certain conditions. White speaks of the role of laws in that kind of inference, but he does not bring the full power of inference analysis to the understanding of the inference relation.

Richard Boyd, though arguing against the inference analysis of cause and effect, describes that view as:

> The version of Hume's account that prevails in twentieth-century empiricist philosophy is [roughly] ... that an event $e_1$ causes an event $e_2$ just in case there are natural laws $L$ and statements $C$ describing conditions antecedent to $e_1$ such that from $L$ and $C$, together with a statement reporting the occurrence of $e_1$, a statement describing the subsequent occurrence of $e_2$ can be deduced.[103]

---

[99] *The Logic of Scientific Discovery*, pp. 59–60. See the discussion of causal explanations in the essay "Explanations" here.
[100] "General laws in history," pp. 233–234.
[101] *Foundations of Historical Knowledge*, p. 60.
[102] Ibid., p. 58 and p. 62.

This agrees with my interpretation of the authors he cites (Hempel and Oppenheim, Feigl, Popper).

Finally, Norwood Russell Hanson says:

> Causes certainly are connected with effects; but this is because our theories connect them, not because the world is held together by cosmic glue. The world *may* be glued together by imponderables, but that is irrelevant for understanding causal explanation. The notions "the cause $x$" and "the effect $y$" are intelligible only against a pattern of theory, namely one which puts guarantees on inferences from $x$ to $y$. Such guarantees distinguish truly causal sequences from mere coincidence.
>
> The necessity sometimes associated with event-pairs construed as cause and effect is really that obtaining between premises and conclusions in theories which guarantee inferences from the one event to the other.[104]

But Hanson does not develop this idea beyond these comments.

Viewing the relation of cause to effect as an inference has been implicit in much of contemporary analyses. But a full development of that idea has been lacking.

---

[103] "Observations, explanatory power, and simplicity: Toward a non-Humean account," p. 355. The authors he cites are on pp. 358–359.

[104] *Patterns of Discovery*, p. 64 and p. 90.

# The Directedness of Emotions

Is every emotion we feel directed at something? Examples from ordinary life suggest not. We can better understand emotions and why we sometimes do and sometimes do not feel justified in calling them directed by using the methods of analysis for reasoning about cause and effect.

Do emotions have to be directed? That is, if a person has an emotion, need it be directed towards something or someone?

Is it possible to have the emotion of fear without being fearful of something or someone? Is it possible to be angry without being angry at someone or something?

By emotion I am going to mean some very strong feeling a person experiences that he, or she, or we, classify as, for example, lust, disgust, rage, grief, fear, awe, joy, or courage. I suspect what I say should generalize to other very strongly felt states that others classify as emotions.[1]

I used to do exercises when I was an actor. They were meant to make me more open emotionally when going on stage. When I first started learning those exercises I was asked to conjure up some image or idea that made me feel one of these emotions and let myself go, then note what my body did. I noted what sounds I made, I noted what my voice was like, and I noted how my body took on roughly the same posture with the same gestures each time. In subsequent lessons I found that I could put on the voice and "mask" of the emotion and, surprisingly to me, the emotion then followed. When I became practiced at this, I found that I could put on the voice and mask of, for example, rage and then feel rage, without any specific image or idea of a person or thing that I was angry at.[2]

---

[1] I am not trying to characterize what an emotion is in this essay. It is enough that you agree that the examples I give are indeed examples of people having emotions. It is not necessary that you agree that we can classify such feelings under particular labels such as "fear" or "anger," though it is helpful in the discussion here that we often can. By "strongly felt" I mean that with the feeling there is also a major physiological change, as I describe below in an example. See Daniel M. Farrell, "Recent work on emotions" for a summary of various views in relation to what I am saying here.

One day in the early 1980s I was sitting in a class about speech acts which John Searle was giving. He said that every emotion has to be directed. I went up to him after class and told him about those acting exercises. His response was that I wasn't really feeling anger.

If you agree with John Searle, then my original question is vacuous. It's like asking, "Is a dog a canine?" If part of what it is to have an emotion is to have it directed, and all other strong feelings are not emotions, then the question is answered.

I do not agree. I think the question is a substantive one, and an answer to it has to be defended. I know that when I did those exercises I felt rage, because I know what rage feels like, and that is about all I think there is to my knowing whether a feeling I am experiencing is rage. It was not an ersatz rage, unless you beg the question and say that any feeling which is not directed towards something cannot be rage. If I am feeling rage, you will be able to see clearly that I am feeling rage—not some pallid version of anger that I am able to conceal, but real rage, for real rage does have physical manifestations. That is not to say that a good actor cannot simulate rage while not feeling it. But if both the person having the experience and those outside classify the feeling as rage, we are fully justified in calling it rage. And then the question of whether the rage was directed at someone or something is not vacuous.[3]

My experiences with the acting exercises convince me that the answer to our question is "No." But you probably have not had such an experience, so I'll give an example that is more common.

Hubert is a moderately good and passionate golfer. He has a bad temper, a short fuse, we'd say. Last Wednesday he went to his country club for a round of golf. On the 7th hole he missed a putt and went berserk. He got really, really angry, he screamed, he cursed, he broke his putter over his knee and threw it in the fairway, and then stomped off to the clubhouse. His face became bright red, his eyes were wide,

---

[2] Compare the training for meditation described in A. Lutz, L.L. Greischer, N.B. Rawlings, M. Ricard, and R. Davidson, "Long-term meditators self-induce high amplitude gamma synchrony during mental practice."

[3] Some people wish to classify emotions as rational or irrational, and perhaps what I am saying here is applicable only to what they would classify as irrational emotions. However, a method or criterion for classifying emotional responses as rational or irrational has not been given; see "Rationality" in *The Fundamentals of Argument Analysis*, a companion volume in this series.

he broke into a sweat, and was trembling. Everyone who was around him said he was in a rage. Later, he said he was in a rage.

What was Hubert angry at? We might say he was angry at himself. But Hubert denies that. If you had stopped him while he was angry and could have gotten him to calm down enough to talk, he might have come up with an answer, but you and he know that when he was overwhelmed by his anger he wasn't really angry at himself or at the ball or at his golf club. He was just angry, in a rage. We describe it by saying that he was *taken over by* his anger, he was *in* a rage. That common way of talking suggests a lack of focus, a lack of direction to the emotion.

Hubert wasn't angry *at* anything at all, unless you want to go on a reifying binge and call "missing his putt" a thing, but even then it isn't right to say he was mad at that.[4] Rather, "Hubert was angry because he missed his putt." Some experiences of rage are not directed, though they have causes.

Compare: "The cat caused Spot to bark." To take that literally is to say that the cat is the *thing* that caused Spot to bark. That way of

---

[4] This is exactly what Peter Goldie does in "Emotions, feelings and intentionality":

> When an emotion is directed towards its object, then this is a sort of *feeling towards* the object. The object can be a thing or a person, a state of affairs, or an action or event: when you fear the lion, the object of your fear is a thing, which has certain determinate properties (sharp teeth, perhaps), that will explain why you fear it; when you are angry about the level of unemployment, the object is a state of affairs (or a fact); and when you are upset at the way she intentionally turned her back on you when you came into the room, the object of your emotion is an action.

If we can stretch the notion of an object this far, then it is not hard to accept that every emotion has an object. But it is highly controversial to accept that states of affairs, actions, and events are things. In my *Predicate Logic* I argue that the only clear characterization we have of the notion of a thing is that which can be reasoned about in predicate logic. I have shown in that book, and in the discussion in an appendix on events in "Reasoning about Cause and Effect" in this volume, as well as in an appendix to my *The Internal Structure of Predicates and Names with an Analysis of Reasoning about Process* that we cannot (easily) reason about events in predicate logic. Similar arguments apply even more so to the notions of states of affairs and actions. We "pick out" actions, events, and states of affairs with propositions (claims), which connects to the causal analysis I present here.

understanding cause and effect is to take inherent powers or forces in things to be the genesis of effects. In "Reasoning about Cause and Effect," I point out that such a view of cause and effect has been abandoned by almost everyone and leaves little guidance for how to reason about cause and effect.

But it is not the cat itself, its mere existence, that caused Spot to bark. It's where it was and what it was doing or not doing that caused Spot to bark. We can more clearly describe the cause as "The cat was near Spot and hissed at him" and the effect as "Spot barked." We use claims to describe purported cause and purported effect, and then the relation of cause and effect can be reasoned about by saying that "The cat caused Spot to bark" is true if and only if "The cat was near Spot and hissed at him *therefore* Spot barked" is a good causal inference. What counts as a good causal inference depends on the metaphysics we adopt, but there is much that is common to all such analyses, as I describe in "Reasoning about Cause and Effect."

To ask what is *the* cause is to ask what thing was a cause, and it is very difficult to explicate that, even if we feel it is somehow right, based on our archetypal examples of cause in which something or someone acts on or towards another. We can get clear about cause and effect, or at least clear enough to be able to reason well about cause and effect, only by identifying the cause with a description we give of a way the world was at a particular time, and identifying the effect with a description we give of a way the world was at a particular time.

Often there is some particular object in such a description that we single out and think of as the cause, like the cat in the example above. But there are many other things in the world that are needed for that description of "the cause" to be true. We pick out one object we deem most significant. To call one of all the things in the world at the time of the true description of the world that we give as being "the cause" seems to be subjective, but often it is intersubjective. Why we have intersubjective agreement on the cat being the thing to pick out may be due to our common backgrounds directing our attention to it, or to the common construction of our bodies directing our attention to it, or because we have some non-sensible apprehension of the real power of causation in the cat. No matter, we often classify a particular thing as the cause.

But often there is no thing we classify as the cause. "The drought caused the crops to fail" could be true, but to say that the drought is a thing is to extend the notion of thing beyond any we can reason with well. Causes, for clear reasoning, are what we describe with a claim about the state of the world.

A similar analysis applies to claims about people having emotions. We can say, "Hubert was angry" and understand that and classify it, in the example above, as true. We can say that "Hubert was angry at something," and to the extent we understand that, we'd say in the example above that it is false. Rather, what is true is "Hubert was angry *because* he missed his putt." "Because"-statements, at least in this context, are "cause"-statements reversed:

"Hubert was angry because he missed his putt" is true
≡ "Hubert missing his putt caused him to be angry" is true
≡ "Hubert missed his putt *therefore* Hubert was angry" is a good causal inference.

When we find it difficult if not impossible to single out a thing or person towards which an emotion is directed, and even in cases where there is a such a thing, such an analysis is clarifying.[5]

*Example 1*  Dick is angry at Zoe.
*Analysis*  It isn't the existence of Zoe, or Zoe as an existing person that Dick is mad at. Rather, filling it out, we have "Dick is angry at Zoe for forgetting to call him." And that reads more naturally as "Dick is angry because Zoe forgot to call him," and recasting the "because" as "cause" we have: "Zoe forgot to call Dick *therefore* Dick was angry" is a good causal inference.

*Example 2*  Lula is overjoyed.
*Analysis*  Lula isn't overjoyed at anything; she's overjoyed because she just found out she passed the examination to become a police officer. That is, "Lula just found out she passed the examination to become a police officer *therefore* Lula is overjoyed" is a good causal inference.

---

[5] I 'm not suggesting that there need be a good causal inference of a particular kind for a strongly felt experience to be classified as an emotion. I am not sure even what it would mean to say that every emotion has a cause; see Example 43, "Everything has a cause," in "Reasoning about Cause and Effect."

The approach I suggest here is quite different from the view that appraisals are causes of emotions, as described in, for example, Rainer Resenzein, "On appraisals as causes of emotions."

*Example 3* Flo is grieving for her doll.

*Analysis* Flo is seven years old and she dropped her doll in the fire and it burned up. She is crying and crying and crying. It surely isn't right to say that she's grieving for her doll, since her doll doesn't exist anymore. Rather, she's grieving for the loss of her doll, or for the burning up of her doll. Clearer is to say "Flo is grieving because she just found out that her doll was burned up," which we can recast by saying that "Flo just found out that her doll was burned up *therefore* Flo is grieving" is a good causal inference.

In many cases of someone having an emotion, there is a thing or person we note in the description of the cause of the emotion that we naturally single out as being the cause of the emotion, or more usually, we say it is that towards which the emotion is directed, using various words like "at" in "angry at," or "for" in "lusting for," or "of" in "fearful of," to signal that directedness. But, as we've seen, there need not be directedness.

When there is directedness, however, it does not seem in any way an arbitrary choice to pick out the object or person in the description of the cause and say that's what the emotion is directed at. When there is directedness, it is because the person who is feeling the emotion picks out that thing or person in the description of the way the world is or was; we, standing outside, can often, though not always, see which object or person it is without being told. Yes, picking out that thing or person is subjective, but that's not like cases of cause and effect where we feel that there should be an objective standard for picking out this thing or that description of the world as the cause. Rather, the person experiencing the emotion having the sensation that the emotion is directed at a particular person or thing is all there is to the emotion being directed at that particular person or thing.

When we feel an emotion directed towards a person or thing, rarely (I hesitate to say never) is it a pure object of the emotion; rather, an object under a description is the object of our emotion. In that case, given a claim about someone having an emotion, we can rewrite it as a "becausal" claim, a claim that can then be rewritten as a causal claim, which can then be rewritten as a claim about a causal inference being good, in which the claim describing the way the world which we identify with the cause includes in some prominent way reference to that person or thing.

# Conditionals

Sentences of the form "if . . . then . . ." play a major role in our reasoning. Some conditionals, as they are called, are claims, and for those we have criteria for when they are true. Some conditionals are intended to be understood as inferences: were this to be true, this would follow. If meant to be judged solely as valid or not, those can sometimes be evaluated by the methods of modal logics. However, we often use conditionals that we deem good that are only strong or moderately strong inferences, and here I present a theory for how to reason with those.

| | |
|---|---|
| Arguments or claims? | 101 |
| The material conditional | 104 |
| The connection between antecedent and consequent in a true conditional | 105 |
| Conditionalization | 106 |
| Strict implication | 107 |
| Strict implication for other notions of necessity | 108 |
| The scope and limitations of modal logic | 110 |
| Conditional inferences | 110 |
| Evaluating conditional inferences | 112 |
| Conclusion | 120 |
| Appendix: Other views of conditional inferences | 121 |

## Arguments or claims?

I give my students an argument:

> Juney is a dog, so Juney barks.

They recognize that something is lacking for it to be a good argument, some claim linking the premises to the conclusion. If I suggest "Almost all dogs bark," they're happy. But if I suggest "If Juney is a dog, then Juney barks" they won't accept the repair. They say that I've just restated the argument. Are they right? Is an "if . . . then . . ." sentence a claim that can be added to an argument as a premise or is it itself a kind of inference?

Consider:

(1) If Caesar had not crossed the Rubicon, then there would not have been a civil war in Rome.

The reasoning we go through in trying to decide if this is true is similar to what we do in analyzing an inference: we consider possibilities and evaluate it on something like the strong–weak scale.

In comparison, consider what the workmen said when they were putting up a gate at my home:

> If this gate was made the right size, then there won't be a gap between it and the post on this side.

Here it doesn't seem to be a matter of possibilities. Either the gate was made the right size or it wasn't; either there is a gap or there isn't. The claim is false if the gate was made properly and yet doesn't fit; otherwise the claim is true.

---

**Conditionals**  A *conditional* is any sentence that is equivalent to one in the form "if . . . then . . .", where "equivalent" means that for all possibilities one of the following holds:

> The "if . . . then . . ." sentence and the original should be evaluated as claims with the same truth-value.

> The "if . . . then . . ." sentence and the original should be evaluated the same as inferences.

The part following "if " must be a claim (allowing for changing the tense) and is called the ***antecedent***. The part following "then" must be a claim and is called the ***consequent***.

---

For example,

> If Ralph is a bachelor, then Ralph is a man.
>
> I'll wash the dishes, if you cook dinner.
>
> Bring me an ice cream cone and I'll be happy.
>
> Being a good husband means taking out the trash.
>
> Either we'll go to the beach or we'll see a movie tonight.

The first is clearly a conditional, with antecedent "Ralph is a bachelor" and consequent "Ralph is a man."

The second can be rewritten as "If you cook dinner, then I'll wash the dishes," and so is a conditional with antecedent "You cook dinner" and consequent "I'll wash the dishes." We can assume that "I" and "you" refer to particular people in the context in which this statement was uttered.

The third can be rewritten as "If you bring me an ice cream cone, then I'll be happy." The antecedent is "You bring me an ice cream cone" and the consequent is "I'll be happy."[1]

The next can be rewritten as "If someone is a good husband, then he takes out the trash," with antecedent "Someone is a good husband" and consequent "He takes out the trash." But who is this "someone"? "He" doesn't refer to a particular person, but to that "someone." How is this a conditional, since the parts of a conditional must be claims? We can rewrite the sentence as:

For anyone, if he is a good husband, then he takes out the trash.

The part "if he is a good husband, then he takes out the trash" is an *open* sentence, one that becomes a claim when we establish to whom "he" refers; part of how we evaluate this claim involves all the ways we could do that. Quantified sentences such as this one can be considered conditionals too, *generalized conditionals*.[2]

---

[1] David Lewis ("Probabilities of conditionals and conditional probabilities," pp. 306–307) apparently believes that the use of "and" requires that a claim such as this one should be analyzed as a conjunction:

> We are gathering mushrooms; I say to you [*] "You won't eat that one and live." A dirty trick: I thought that one was safe and especially delicious, I wanted it myself, so I hoped to dissuade you from taking it without actually lying. I thought it highly probable that my trick would work, that you would not eat the mushroom, and therefore that I would turn out to have told the truth. But though what I said had a high subjective probability of truth, it had a low assertability and it was a misdeed to assert it. Its assertability goes not just by probability but by the resultant of that and a correction term to take account of the pointlessness and misleadingness of denying a conjunction when one believes it false predominantly because of disbelieving one conjunct.

Frank Jackson in *Conditionals*, p. 25, also considers [*] to be a conjunction. I think [*] is the negation of a true conditional (not: If you eat that one, you'll live), given Lewis' story.

[2] See my *Predicate Logic*, especially p. 47 and pp. 100–102.

It is not clear whether we should take the final example as a conditional. Some say it is equivalent to "If we don't go to the beach, then we'll go to a movie tonight." Claiming an equivalence between "if . . . then . . ." claims and "or" claims is quite common, and if we agree to that in this case, then we can classify the sentence as a conditional.

Often conditionals have a clause in the subjunctive to show that we are talking of possibilities about the corresponding non-subjunctive claims. We allow for such rewriting in saying that the antecedent and consequent of (1) are claims: "Caesar did not cross the Rubicon" and "There was not a civil war in Rome."

## The material conditional

If a conditional is not an inference, how do we evaluate it? Philo of Megara gave the following condition.

---

***The material conditional*** A conditional is true whenever it is not the case that its antecedent is true and consequent false.

---

In tabular form:

| A | B | If A then B |
|---|---|---|
| T | T | T |
| T | F | F |
| F | T | T |
| F | F | T |

The first two lines are determined because we want to use conditionals in our reasoning to argue by ***modus ponens***: If A then B; A ; so B. The last two lines are best illustrated with a generalized conditional:

(2) For every $x$ and y, if $x$ and $y$ are married, then $x$ and $y$ can file their income taxes jointly.

This is true (in the United States). But if it is true, it must be true no matter who $x$ and $y$ refer to. So "If Maria and Tom are married, then Maria and Tom can file their income taxes jointly" must be true, though Maria and Tom aren't married and they can't file their taxes jointly. So with false antecedent and false consequent, the claim is true. But also "If John and Horace are married, then John and Horace can file their taxes jointly" must be true, where John and Horace aren't married but have entered into a domestic partnership recognized by law that allows them to jointly file their taxes. So a claim with false antecedent and true consequent must be true. The material conditional allows us to classify sentences such as (2) as true by ignoring what happens when the antecedent is false.

In a full analysis of reasoning with claims as wholes called *classical propositional logic* the key assumption that generates this evaluation of conditionals is the *classical abstraction*: The only properties of a claim that matter to logic are its form and its truth-value. Where that assumption fits the context of an argument, Philo's analysis of conditionals is the best choice.

## The connection between antecedent and consequent in a true conditional

All other criteria that have been proposed for conditions under which a conditional is true assume that some other aspect of claims has to be considered in reasoning: how claims could be true or false, their subject matter, how we could come to know they are true, how likely it is that they are true, . . . . One particular aspect is fixed on, and we then ask whether the antecedent and consequent are appropriately related. The general form of evaluation is the following.

---

***Conditions for a conditional to be true*** A conditional "If A then B" is true if and only if both:

It is not the case that A is true and B is false.

The aspect of A other than its truth-value and form that is under consideration is appropriately connected to that same aspect of B.

---

That is, the conditions for the material conditional must hold, and the aspect of claims being paid attention to must be related in the way considered significant for that analysis.

For example, intuitionists take the aspect that is of concern to be the ways in which we could, theoretically, come to know that a claim is true; the relation is that every way in which we could come to know the antecedent is a way we could come to know the consequent as well.

One ancient Greek analysis says that a conditional is true if its consequent is included in its antecedent. In a modern version, dependence logic, this is formalized by taking the aspect of a claim to be its referential content: what objects are referred to and/or predicates used in the claim. The connection is that the referential content of the antecedent must include that of the consequent. In other formalizations the content of the antecedent must be included in that of the consequent. Or in subject matter relatedness logic, the contents must overlap.

Each of these analyses fits into a general overview of propositional logics: logical systems formalizing how to reason with claims as wholes, ignoring their internal structure except for the use of "and," "or," "not," and "if . . . then . . .". Each seems entirely appropriate for reasoning when the particular aspect that generates the logic is of concern; each seems inappropriate when that aspect is not what is important to us in our reasoning.[3]

## Conditionalization

In classical propositional logic there is a clear relation between conditionals and arguments. A *tautology* is a claim that is true for every possible way the world could be. For example, each of the following when interpreted in classical propositional logic is a tautology:

Juney is a dog or Juney is not a dog.

If Juney is a dog, then Juney is a dog.

Not both Juney is a dog and Juney is not a dog.

If a material conditional is a tautology, it is certainly true. But more:

It is not possible for both the antecedent to be true and consequent false at the same time.

Reading "premises" for "antecedent" and "conclusion" for "consequent," this is the condition for an inference to be valid. For the material conditional, in the context of the analysis of classical propositional logic, the following rule holds.

---

**Conditionalization of an inference**  The *conditionalization of an inference* "$A_1, A_2, \ldots, A_n$, therefore B" is "If $A_1$ and $A_2$ and. . . and $A_n$, then B."

**The Deduction Theorem**  An inference is valid if and only if its conditionalization is a tautology.

---

[3] See my *Propositional Logics* for the development of this view.

For some of the logics discussed in the previous section, the Deduction Theorem also holds.[4] For others, it does not.

## Strict implication

One analysis of conditionals is to view them as inferences judged solely for validity.

---

***Strict implication*** A conditional is evaluated as a *strict implication* means that "If A, then B" is true if and only if there is no way in which the antecedent could be true and the consequent false.

---

The truth-conditions of conditionals evaluated as strict implications are those for the inference "A, therefore B" to be valid. Hence the name *implication* or *entailment* for conditionals that satisfy this condition.

On the strict implication reading, "If Dick arrived at the restaurant after 6 p.m., then he worked late" is false: he might have arrived late due to a hail storm or a car accident.

This reading of conditionals is normally not meant to replace other readings but to supplement them. To make the discussion simpler, I'll discuss just classical propositional logic, where the usual formalizations of "and," "or," and "not" are given along with "if . . . then . . ." as the material conditional. Added to those is the option to read some conditionals as strict implications. Then a claim is defined to be *necessary* if it is true for every possible way the world could be; a claim is *possible* if there is some way it could be true. Usually one of these notions is taken as primitive in the language, and then strict implication is defined via:

The strict implication "If A, then B" is true if and only if the material conditional "If A, then B" is necessary.

That is,

Necessarily: If A, then$_{(material\ conditional)}$ B.

 is equivalent to: If A, then$_{(strict)}$ B .

The words "it is possible that" (or "possibly") and "it is necessary that" (or "necessarily") are *modal* words/phrases: they show the mode, the way in which a claim is to be taken as true or false. The

---

[4] See my *Propositional Logics*.

logic of strict implication built on the classical propositional logic is called *classical modal logic* or the classical modal logic of *logical necessity*.

This logic can be used to formalize and guide us in our analyses of whether an inference in classical propositional logic is valid, as well as to clarify the relation of various inferences in classical propositional logic. Classical modal logic formalizes the rule of conditionalization, clarifying exactly how we shall look at "all possible ways" in deciding if an inference is valid.

## Strict implication for other notions of necessity

The modal logic approach to formalizing how we investigate whether an inference is valid has been extended to other notions of possibility and necessity.

*Physical necessity* corresponds roughly to how we might reason in science. A claim is necessary if it is true in every possible way that shares the same physical laws as our world. There is, apparently, nothing inconsistent in assuming that the power of gravity is proportional to the inverse cubed instead of the inverse squared of the distance between two objects; that is a logical possibility but not a physical possibility relative to our current science. Nonetheless, we might wish to analyze what would happen in such a possibility. Relative to that description in which gravity acts differently than here, we could ask what is physically necessary.

*Time-dependent necessity* or *inevitability at time t* corresponds to our asking whether a claim is necessary given how things are at a particular time: a claim is necessary if it is not possible that it *will be* false. Is it necessary right now that Caesar crossed the Rubicon? Yes, since that is how the world is (was). But we can consider how the future might be: If a transvestite is elected president of the U.S. in 2040, is it necessary that there be other political parties than there now are in the U.S.? We look at all possible ways things could be relative to the description in which a transvestite is president.

The logics of physical necessity or time-dependent necessity, built on classical propositional logic or on some other propositional logic, give us some insight into the relation of inferences and conditionals and of what forms of claims are necessary. But in order to use physical necessity in the analysis of any particular claim or inference, the physical laws that are assumed to hold must be made explicit, just as we need to make the premises of any inference explicit. In order to use time-dependent necessity, we must say exactly which claims we are assuming are true about the way the world is now, just as we need to make the premises of any inference explicit when challenged. And that is exactly what is glossed over.[5] A formal tool is invoked, but the hard work that makes it useful is "assumed," as if all of us can imagine very clearly what explicit premises are required. It is not sloppiness or laziness that makes practitioners of these approaches do that. It is a faith that there are such real things as possible worlds, we just can't describe them very well.

But what we need of possible worlds in our reasoning is that they correspond to (can be represented as) consistent collections of claims. And we must have that, for the claims assumed to be true in a possible world are just the premises of our inferences. We leave out much in

---

[5] See the appendix to "On valid inferences" in *Reasoning and Formal Logic* in this series. A similar point is made by Mates in *The Philosophy of Leibniz* where he discusses Leibniz's view that nothing could be different than it is. Could Caesar still have been the same person and not have crossed the Rubicon? Some say yes, because we stipulate possible worlds, and we can stipulate that the individual Caesar in the world we are discussing didn't cross the Rubicon. But, says Mates,

> We assume that everything else in [that possible world] in which that is the case is pretty much the same as in this world, except for such facts as are inconsistent with the hypothesis about Caesar's crossing. In particular, I suppose, we assume that the laws of biology and physical science generally continue to hold.
>
> These kinds of assumptions alone, as Leibniz pointed out, threaten to give us quite a mess, and matters become still more complicated when we assume in addition that, on the whole, the same individuals exist in the posited world as in ours. After all, the event called "Caesar's crossing of the Rubicon" was the result of myriad antecedent conditions and itself led to a limitless ramification of consequences. The great preponderance of these antecedents and consequences are completely beyond our ken, of course, though presumably knowable in principle. If we did know the whole story, however, we might even find that under the given conditions it was impossible that Caesar did *not* cross the Rubicon; perhaps for the crossing not to happen there would have had to occur certain antecedent circumstances that in turn would imply, by the relevant physical and biological laws, that it would have to happen, after all. p. 252

how we reason, but we must be prepared to fill in the gaps. Scientists, for example, make explicit what physical laws they assume in deriving their conclusions.

## The scope and limitations of modal logic

The strict implication analysis of conditionals in modal logic has several values:

- It formalizes how to evaluate whether an inference is valid.
- It allows us to investigate the rule of conditionalization.
- It gives us a formal framework in which to consider issues of possibility and necessity.
- It allows us to compare these points for various metaphysics codified in different propositional logics.

But there are serious limitations to the use of modal logic. Great care must be taken in using it since it requires us to say that "If Caesar had not crossed the Rubicon, then there would not have been a civil war in Rome" is false, where that means it is (corresponds to) an invalid inference. The important distinction between the true/false division and the valid/strong–weak scale can be easily confused.[6]

In any case, my students never thought that "If Juney is a dog, then Juney barks" was a condensed valid argument. Modal logics formalize the rule of conditionalization, but they ignore the possibility that a conditional could be intended as an inference lying on the scale from strong to weak.

## Conditional inferences

Consider the following conditionals:

---

[6] Some logicians think that all modal logics involve a use-mention mistake: Iin a strict implication, it is said, we are talking *about* the antecedent and consequent, not using those claims. See, for example, Dana Scott, "On engendering an illusion of understanding."

There is also a difficulty with self-referential paradoxes, since "possible" means "possibly true"; see Example 18 of Chapter VI of my *Propositional Logics*. In *The Internal Structure of Predicates and Names with an Analysis of Reasoning about Process* I discuss the view of "possibly" and "necessary" as adverbs, modifiers of the predicate "— is true."

(3) If that piece of butter had been heated to 30°C, it would have melted.

If the Norsemen did settle in America, they didn't stay there long.

If he's out in this storm, he'll be soaked.

If I had known that the foundation of this house was cracked, I would not have bought the house.

Apply for that job and maybe you'll get it.

All of these are conditionals. But none of the analyses of conditionals we have looked at seem appropriate. We want to consider possibilities when we evaluate them, yet they don't seem to be intended as valid inferences.

All of these are best understood as abbreviated inferences:

That piece of butter was heated to 30°C *therefore* it melted.

The Norsemen settled in America *therefore* they didn't stay long.

He is out in this storm *therefore* he will be soaked.

I knew that the foundation of this house was cracked *therefore* I did not buy the house.

You apply for that job *therefore* you get the job.

These sound odd because we are used to using "therefore" in the sense of an argument. Put "suppose" before the premise and replace "therefore" with "then," and they sound right:

Suppose that piece of butter was heated to 30°C. Then it melted.

Suppose the Norsemen settled in America. Then they didn't stay long.

Suppose he is out in this storm. Then he will be soaked.

Suppose I knew that the foundation of this house was cracked. Then I would not have bought the house.

Suppose you apply for that job. Then you will get the job.

Conditionals like those at (3) are used for us to suppose something, leaving open whether the antecedent is true or false; they are **conditional inferences**. In these examples, "then" is used as "therefore," but not the "therefore" of arguments. These are

*112    Conditionals*

inferences used for reasoning by hypothesis, where we do not assume or require that the premise be true for the inference to be good.[7]

As further evidence that conditionals such as these are best understood as inferences, note that we can add indicator words to the consequent such as "maybe," "probably," and "most likely." For example,

> If the butter had been heated to 30° C, it would certainly have melted.
>
> If I had known that the foundation of the house was cracked, I probably wouldn't have bought the house.
>
> If you get that job, you'll most likely find that you'll like it.

These indicator words are not part of the consequents. They tell us where on the scale from valid to weak the speaker believes the inference lies.

## Evaluating conditional inferences
Consider:

> If I had known that the foundation of the house was cracked, I probably wouldn't have bought the house.

As an inference this is:

> I knew that the foundation of the house was cracked.
> *Therefore*, I did not buy the house.

The word "probably" shows that the speaker intends the inference to be judged as moderately strong, but not very strong, certainly not valid. Even then, the one premise is not enough to get the conclusion. Premises are missing. What do we add?

> Everything else was just as it was at the time of the purchase of the house.
>
> I knew nothing more about the house than I did then, except that the foundation was cracked.

---

[7] What I call "conditional inferences" include conditionals that have been studied under other names. *Counterfactuals* are conditionals whose antecedent is clearly false or dubious (not ones whose antecedent simply is false, because we use many of those as claims in reasoning by *modus tollens* in classical logic). Others are *possible but improbable* conditionals. *Subjunctive* conditionals are ones in which the verbs are in the subjunctive mood.

But these are not enough. It's not only that things were just as before except for this one claim. The premise that the speaker knew that the foundation was cracked must be assumed not to have changed what would have happened except for the speaker making the decision to buy the house. Perhaps we should add:

> The bank would have lent money for the house despite my knowing that the foundation was cracked.
>
> I would still have trusted the contractor I hired to inspect the house knowing that the foundation was cracked.

But is this right? Shouldn't the speaker's knowing that the foundation was cracked be assumed to throw doubt on his or her confidence in the inspector, and hence be a factor in whether the speaker bought the house?

We don't know. We can't use the usual rules and methods for repairing inferences that we have for arguments because those assume that the speaker has good reason to believe the premises are true. What, exactly, does the speaker think is plausible relative to a claim he or she knows is false? There is absolutely no way for us to know. *We have no guide to adding premises to a conditional inference.*

There are, however, certain kinds of conditional inferences for which it is more or less clear what premises are, or need to be, assumed. Consider:

(4) If that piece of butter had been heated to $30^{\circ}$C, it would have melted.

This would be like physical necessity: we assume the physical laws we know hold when the butter is heated to $30^{\circ}$C. The inference is valid relative to those. Compare:

> If that piece of butter had been heated to $8^{\circ}$C, it would have melted.

This is a weak inference, relative to the physical laws we know. If the speaker had intended that we assume other physical laws, he or she should say so.

Between these extremes there are many inferences for which we can assume some background. Consider:

(5) If the light had been red, I would have stopped and there would have been no accident.

The speaker believes "the light was not red" when she entered the intersection. She shows that by a *reductio ad absurdum* inference, deriving that there would have been no accident if the light had been red. Here physical laws are assumed, as well as some psychological assumptions, which are not very clear. Do we have to assume further that nothing extraordinary would have occurred?

> My brakes did not fail.
>
> No car hit me from behind.
>
> A great gust of wind did not push me across the intersection.
>
> The operator of the large crane down the block did not lose control of it and let the ball swing loose and hit my car from behind.

The premises we have to add are what the speaker considers to be the *normal conditions*. Some of them are obvious and can be left unsaid. When the strength of the conditional as an inference depends only on obvious unstated premises, as with (4) or (5), we can reasonably evaluate it.[8]

It might seem that the rule of conditionalization for conditional inferences should be: A conditional inference is good if and only if the corresponding inference is valid or strong. But that's wrong. A conditional inference might not be intended as valid or strong:

> If you get that job, you might find that you'll like it.
>
> If I had known that the foundation of the house was cracked, I probably wouldn't have bought the house.
>
> If winter comes early, I most likely will go to Costa Rica for Christmas.

The first should be judged as good if the corresponding inference is even the least bit strong. The second is good if the inference is moderately strong. The last—well it's not really clear, but it should be stronger than the first conditional inference, though not very strong.

---

**Good conditional inferences**  A conditional inference is *good* if and only if the corresponding inference has the strength indicated in the conditional relative to the obvious unstated premises.

---

[8] Compare the discussion of normal conditions for causal inferences and of hypothetical causal claims in "Reasoning about Cause and Effect" in this book.

It follows that:

- If it's not clear what further premises need to be assumed, the conditional inference is not good.
- If it's not clear how strong the speaker thinks the inference is, the conditional inference is not good.

Just as with an argument that is lacking too much, we judge a conditional inference bad if the speaker left out too much. As with analogies, most conditional inferences are not good, since too much is left out; they are best thought of as suggestions for hypothetical arguments.

So in a sense my students were right. The conditional "If Juney is a dog, then Juney barks" can be understood in that context as a restatement of the argument "Juney is a dog, so Juney barks." On that understanding, it is good only if it is valid or strong, since that is the strength required for arguments.

Conditional inferences are good or bad, not true or false. Yet since they seem to be ordinary declarative sentences, some say they are true or false. Compare strict implications. It is harmless to think of a conditional inference as a claim and call it true or false if by that we just mean that it satisfies the conditions for a good conditional inference and if we do not then fall into use-mention confusions. Words like "probably" and "most likely" are not part of the antecedent or consequent, yet they are essential in conditional inferences in telling us how to judge whether a conditional is good.

*Example 1* The defendant is guilty of shoplifting. The officer saw him pick up the watch, put it on his wrist, and walk out of the store without paying. (*) If he had intended to pay, then almost certainly he wouldn't have done all that.

*Analysis* This is an argument. The premises are:

The officer saw the defendant pick up the watch.

The officer saw the defendant put the watch on his wrist.

The officer saw the defendant leave the store with the watch without paying.

If the defendant had intended to pay, then almost certainly he wouldn't have done all that.

The conclusion "The defendant is guilty of shoplifting" then follows by *modus tollens* (If A, then B; not B; so not A) on (*).

*116   Conditionals*

But why should *modus tollens* apply to conditional inferences? *Modus tollens* can be justified for conditionals that are intended to be and are valid or strong: that's reasoning by *reductio ad absurdum*. In this case we have:

(a) Suppose the defendant intended to pay.
(b) Then he would not have done all that.
   He did do all that.
   Therefore he did not intend to pay.

Now it is clear that some premises are missing, ones to make the inference from (a) to (b) strong. The speaker is under an obligation to provide those if he or she wants us to accept the conclusion.

*Example 2*   The defendant is guilty of shoplifting. The officer saw him pick up the watch, put it on his wrist, and walk out of the store without paying. (‡) If he had intended to pay, then there's a good chance that he wouldn't have done all that.

*Analysis*   Viewing (‡) as a conditional inference, the argument would be the same as in Example 1 except the inference from (a) to (b) there is assumed to be only a bit strong, so unstated premises are not called for to make it very strong. In that case, the argument is too weak to justify the conclusion: the analogue of *modus tollens* for conditional inferences does not go through in this case.

*Example 3*

Manuel is handicapped. (\*\*) If Manuel is going fast in his wheelchair past a finish line that has a banner with the Olympic rings and people are lined up cheering, then Manuel is almost always racing in the Olympics for the handicapped. Manuel is going fast in his wheelchair past a finish line that has a banner with the Olympic rings, and people are lined up cheering. So Manuel is racing in the Olympics for the handicapped.

*Analysis* This is an argument submitted by one of my students who was asked to prove the conclusion based on the cartoon, where Manuel is the person in the wheelchair.

My student knew that a valid argument here would need a dubious premise, so he wished to make his argument only strong. He knew that "Almost all dogs bark" is a better premise than "All dogs bark," so he used the claim (\*\*) which has the form "If . . . then almost always . . .". How should we understand that?

We could simply delete "almost always," viewing those as indicator words showing that he thought the claim was not certainly true. Then the argument is valid, but it has an uncertain premise.

Or we could view (\*\*) as a conditional inference, with the words "almost always" indicating that the student thought it only strong. Then we have a strong argument with clearly true premises.

*Example 4* Manuel is handicapped. (‡‡) If someone is going fast in his wheelchair past a finish line that has a banner with the Olympic rings, and people are lined up cheering, then he is almost always racing in the Olympics for the handicapped. Manuel is going fast in his wheelchair past a finish line that has a banner with the Olympic rings, and people are lined up cheering. So Manuel is racing in the Olympics for the handicapped.

*Analysis* Here the conditional is open: "someone" and "he" must be quantified to make the "if . . . then . . ." sentence (or parts of it) into a claim. The usual assumption in ordinary speech is that the sentence is universally quantified: "for anyone, if he is going fast . . .". On that reading, we can take "almost always" as a commentary on the plausibility of the universally quantified conditional.

Or we could view "almost always" as an odd way of saying "for almost all things": a nearly universal quantifier used in place of the universal quantifier. Then claim (‡‡) is plausible and we have a good strong argument.

These are sophisticated readings of the argument which I doubt my student could have distinguished. But then, he didn't present this argument; he gave the one in Example 3.

*Example 5* If Spot wouldn't chase cats, then he would be sick. If Spot were a cat, then he wouldn't chase cats. So if Spot were a cat, he would be sick.

118   Conditionals

*Analysis*   There's something wrong in this reasoning about Spot the dog. We might accept the first two conditionals, but the third one seems wrong. Aren't conditionals transitive: from "If A, then B" and "If B, then C", can't we conclude "If A then C"?

To even think of the problem of the transitivity of conditional inferences as one that can be resolved in a general way ignores what we've learned about the particularity of the analysis of each conditional inference. Often it's not even clear how to form "If A, then C" given conditional inferences "If A, then B" and "If B, then C." What should be the asserted strength of "If A, then C"? What normal conditions are to be assumed? Even if we can identify those, the normal conditions might not be compatible: what is assumed true to make the inference from A to B good may be denied in making the inference from B to C, as in this example.

*Example 6*   Tom: If Dick ate the last of the pizza in the refrigerator, then he didn't remember that Zoe told him they were going out to dinner.

*Analysis*   We could analyze this as a material conditional: We ask whether Dick ate the last of the pizza. If he did and he also remembered that Zoe told him they were going out to dinner, then the claim is false.

Or we could analyze it in terms of subject matter relatedness. Since the antecedent and consequent have some subject matter in common, the evaluation is reduced to evaluating it as a material conditional.

Or we could analyze it according to the logic of intuitionism. Is every way we could come to know that Dick ate the last of the pizza also a way we could come to know that Dick didn't remember that Zoe told him they were going out to dinner? No. So the antecedent and consequent are not appropriately related, and the conditional is false.

Alternatively, a colleague said:

> It would also be false if Dick did not eat the pizza, but might well have done so even if he remembered what Zoe told him.
> (He thinks he can always stuff in some more.)

This is an examination of possibilities, and so takes the example to be a conditional inference. As an inference, my colleague's story suggests that it is weak. Since Tom put no qualifiers on the inference, it would seem he didn't put it forward as weak, so the conditional inference is bad.

There are many ways to evaluate conditionals. Unless we know how the speaker intends the conditional, possibly inferring that from context, all we can do is present a range of analyses.

*Example 7*  If it's really true that if Dick takes Spot for a walk he'll do the dishes, then Dick won't take Spot for a walk.

*Analysis*  We use conditionals with conditional antecedents in our daily lives. Ignoring the phrase "it's really true that" as just a way to form a compound conditional in English, this example has the form:

If (if A, then B), then not-A.

Without a way to determine how the speaker intends the example to be analyzed, we have to survey the various approaches.

As a material conditional, the antecedent is true, because Zoe will make it so. Thus, the entire conditional is false just in case Dick takes Spot for a walk.

As a subject matter relatedness conditional, since the antecedent and consequent are related, the analysis reduces to that of the material conditional.

As a conditional in intuitionistic logic, we first ask if the antecedent is true. Is it possible for us to come to know that Dick takes Spot for a walk without our knowing that he'll do the dishes? Yes, we might just see them walking down the street. So the antecedent is false. In each of those ways we come to know the antecedent of the whole conditional is false, can we come to know the consequent is false? No. So the conditional is false.

Evaluated as a conditional inference, the example is:

If Dick takes Spot for a walk, then he'll do the dishes.
Therefore, Dick won't take Spot for a walk.

This sounds better. Is it possible for the premise to be true—given what we know of Dick, Zoe, and Spot—and the consequent false? Intuitively, no. The inference is valid or very strong. But how do we evaluate the premise? Should we analyze it as a claim? In those cases in which it comes out true, again relative to what we know of Dick, Zoe, and Spot, "Dick won't take Spot for a walk," I think, will be true, too. But we might consider the antecedent to be a conditional inference. In that case we have:

Suppose Dick takes Spot for a walk.
Then, he'll do the dishes.
So, Dick won't take Spot for a walk.

Since it is very unlikely that Dick will do the dishes, we have a good example of *reductio ad absurdum*.

*Example 8* If that piece of butter had been heated to 8°C, it would have melted.

*Analysis* We've seen how to analyze this as a conditional inference. However, we might consider it to be an assertion of cause and effect were things different than they had been: that piece of butter having been heated to 8°C would have caused it to melt. The evaluation of hypothetical causal claims proceeds just as for a conditional inference except that more conditions are required in order for it to be true.[9]

## Conclusion

We can view some conditionals as claims. Then they are true or false according to whether they satisfy the table for the material conditional as well as, perhaps, having the content of their antecedent and consequent appropriately related. Sometimes we can view conditionals as inferences that are meant to be judged solely as to whether they are valid. Those analyses are codified in the work of modal logics.

But often in our daily lives and in science we use conditionals that are meant to be judged as inferences that are not necessarily valid. For such a conditional to be good we need to be able to fill in what other premises might be needed for the inference and then see whether the inference has the strength that is required of it according to the context in which it is used. If it's not obvious what other premises are needed, or if we cannot tell what strength the inference is meant to have, then the conditional is not good; it is at best a suggestion for reasoning by hypothesis.

Conditional inferences do not satisfy the rules that we expect of conditional claims. They are not in general transitive, nor can we always apply the rule of *modus tollens*. But we can often use them to derive conclusions in arguments if we pay attention to how they are used.

---

[9] See Example 41 of "Reasoning about Cause and Effect" in this volume.

# Appendix: Other views of conditional inferences[10]
## Chisholm and Goodman
For Roderick Chisholm and Nelson Goodman:

> The proposed analysis is this: that any subjunctive or contrary-to-fact conditional, not analytically true, is equivalent to the statement that there is a set of statements, S, taken to be true, which, conjoined with A, entails or strictly implies C. The set of statements, S, is not specified in the assertion of any counterfactual; but anyone making such an assertion implicitly commits himself to holding that there is such a set of statements and that they are true.[11]

This is similar to what I propose in that it takes conditionals which have clearly false antecedents (*counterfactuals*) or are in the subjunctive mood as inferences. But this approach has two problems:

- Many conditional inferences are not intended as valid, and they should not be judged according to that standard.
- A conditional inference may depend on unstated premises, but not necessarily true ones.

Goodman considers primarily the role that counterfactuals play in science. It may be that in the context of scientific inquiry, counterfactuals, such as "If that piece of butter had been heated to $30°C$, it would have melted," are to be understood as corresponding to valid inferences, where the unstated premises are assumed-to-be true physical laws and the way things were at the specified time.[12]

## Rescher

> The outcome of [Nicholas] Rescher's discussion is that there can be no logical resolution of the problem of counterfactuals: it is from

---

[10] For a survey of various approaches to conditional inferences see Chapter 3 of J. L. Mackie's *Truth, Probability and Paradox*. Mackie holds that counterfactuals are condensed inferences, though there are differences in our views.
[11] R. S. Walters, "Contrary-to-fact conditional," p. 213. This is a bit oversimplified, but close enough for the discussion here. See in particular Nelson Goodman's "The problem of counterfactual conditionals" and Chisholm's "Contrary-to-fact conditionals."
[12] But Chisholm in "Law statements and counterfactual inference" argues that there is no good distinction between law and non-law statements. Igal Kvart, *A Theory of Counterfactuals*, Chapter 1, argues that this kind of analysis is susceptible to trivialization by taking the "wrong" true statements or laws. See the discussion of laws in "Reasoning about Cause and Effect."

## 122  Conditionals

> extralogical resources, or the dialectical setting, that resolution comes. ... [A]nyone, including ourselves, asking us to make a belief-contravening supposition must be prepared to interpret and select, when considering the bearing of the supposition on other relevant beliefs. Rescher nevertheless distinguishes between nomological, or law-governed, counterfactuals, on the one hand, and purely hypothetical counterfactuals, on the other. He holds that we can adjudicate, in any case of disputes where the first kind are concerned, between plausible and implausible counterfactuals. In the case of the second, he says that perplexity results if we are asked to resolve a case where counterfactuals seem to compete. We can, for example, distinguish plausible from implausible counterfactuals, each beginning with "If Jones had eaten arsenic." But the distinction is not clear where they are purely hypothetical counterfactuals, each beginning, for example, with "If Hume and Voltaire had been compatriots."[13]

Rescher seems to be trying for some precision in dividing up counterfactuals according to (in my terms) how obvious it is to supply unstated premises. In some cases it's pretty straightforward, but that doesn't seem to depend on laws. Though the plausibility of counterfactuals is considered, thereby suggesting they be evaluated on a continuous scale and not just true/false, the evaluation of a counterfactual is not connected to the strength of the corresponding inference.

### *Stalnaker and Lewis*

Robert Stalnaker sets out the realist's difficulty in analyzing conditionals:

> This problem derives from the belief, which I share with most philosophers writing about this topic, that the formal properties of the conditional function, together with all of the *facts*, may not be sufficient for determining the truth-value of a counterfactual; that is, different truth-valuations of conditional statements may be consistent with a single valuation of all nonconditional statements. The task set by the problem is to find and defend criteria for choosing among these different valuations.[14]

Stalnaker is looking for objective criteria for determining the truth of conditionals, particularly counterfactuals. Noting that conditional inferences involve analyzing possibilities, he devises a formal apparatus of possible worlds for judging whether a counterfactual is true. That requires "selecting" a possible world in which to analyze the corresponding material conditional.

> The world selected *differ[s] minimally* from the actual world. This implies, first, that there are no differences between the actual world

---

[13] R. S. Walters, "Contrary-to-fact conditional," p. 214.

[14] "A theory of conditionals," pp. 98–99.

and the selected world except those that are required, implicitly or explicitly, by the antecedent. Further, it means that among the alternative ways of making the required changes, one must choose one that does the least violence to the correct description of the actual world. These are vague considerations which are largely dependent on pragmatic considerations for their application.

This suggests that there are further rules beyond those set down in the semantics, governing the use of conditional sentences. Such rules are the subject matter of a *pragmatics* of conditionals. Very little can be said, at this point, about pragmatic rules for the use of conditionals.[15]

David Lewis modifies Stalnaker's approach.[16] We put a measure on possible worlds of how similar one possible world is to another in the sense that most of what is true in the one is true in the other. Then a counterfactual "If A, then B" is true if and only if in every possible world that is within the range of possible worlds similar to ones in which A is true, the material conditional "If A, then B" is true.

Neither Stalnaker nor Lewis account for conditionals as inferences that are not intended to be valid or very strong. When I say "If winter comes early, I most likely will go to Costa Rica for Christmas" I do not mean that in every description of the way things could be in which winter comes early, and which are sufficiently similar to that world, the material conditional is true. Rather, in most of those worlds the material conditional is true. To accommodate this into their approach is just to accommodate all of inference analysis in a formal apparatus of possible worlds.

These realist approaches to understanding conditionals give us no guide for how to reason with conditionals. Stalnaker is explicit about this, for he assigns all the work of presenting criteria for justifiably asserting conditionals to "pragmatics." Lewis puts that problem into the objective but unknowable similarity relation.

We need to know the speaker's intent to decide which "worlds" he or she is considering, that is, what further assumptions he or she is making that fill out the assumption that the antecedent would be true. The analysis I give says that if we can't figure out the further unstated premises, it's not a good counterfactual. The focus on objective standards that lead to the classification of every conditional as true or false is misleading. The conditional, in these cases, is a kind of inference, and by my standards the issue is whether we are justified in using it in our reasoning. Perhaps there is an objective standard independent of what the speaker thought, but that will be of little use to us.

---

[15] Ibid., p. 104 and p. 110.
[16] *Counterfactuals.*

## Kvart

Igal Kvart bases his analysis on a notion of objective probability that applies to events.[17] With the world at a certain time, there is an objective probability that the world will be in a particular state at another time. For example, there might be a 50% probability that flipping a coin it will land heads up, independently of anything we may believe or reason about the situation; there might be an 11.26732% chance that Dick will do the dishes after he takes Spot for a walk. Using that notion Kvart defines one event being causally relevant to another if, roughly, the probability of the latter increases conditionally on the former. He then defines the truth-conditions for conditionals by supplying premises for the inference from the antecedent to consequent according to whether they are causally relevant to the consequent, given the antecedent.[18]

Kvart retreats from the unknowable objective probabilities to:

> ... a qualitative interpretation of probability. Notice that a qualitative interpretation can involve the relation "greater than" between conditional probabilities, as well as that of equality.[19]

In practice, that qualitative notion seems to be just logical-relation-probability.[20] Still, he believes that each conditional inference has a truth-value. In *A Theory of Counterfactuals*, p. 234, he considers the example:

> Had Rudolf Carnap been teaching philosophy in Princeton during the 1940s, Princeton would have turned out more logical positivist graduates.
>
> Now there are obviously various hypothetical processes that could have resulted in Carnap's teaching at Princeton in the 1940s. But if the above counterfactual is true, ...

I do not see how the conditional could be true, given no specification of the "hypothetical processes." Compare the example about Voltaire and Hume in the discussion of Rescher above.

## Jackson

Frank Jackson takes pragmatics seriously. In his book *Conditionals* he first divides all conditionals by their grammar: those in the indicative tense vs. those in the subjunctive.[21]

---

[17] *A Theory of Counterfactuals*. See also the appendix on objective chance in "Reasoning about Cause and Effect" in this volume.
[18] See particularly p. 74 of *A Theory of Counterfactuals*.
[19] Ibid., p. 74.
[20] See "Probabilities" in *The Fundamentals of Argument Analysis* in this series for a definition of this and a comparison to other notions of probability.
[21] Jackson does not give a guide for dividing conditionals in languages that do not employ a subjunctive tense.

An indicative conditional "If A, then B," he says, has the truth-conditions of the material conditional. But for us to be epistemically warranted to assert it, the probability of B given A must be high. This is his "supplemented equivalence theory" (p. 37). In addition, for us to use an indicative conditional in our reasoning, the high probability of the conditional must not be affected by A being highly probable. This latter condition is needed "to ensure the utility of Modus Ponens" (p. 29). He explains that the notion of probability he is using is Ramsey's subjective degree of belief (p. 14).[22] Thus, Jackson's condition for "If A, then B" to be assertible is that "A therefore B" be valid or strong, and for us to use such a conditional that evaluation must not be changed by our finding that A is plausible.

Jackson wants to explain the paradoxes of the material conditional (that is, "not-A; therefore if A, then B" and "B; therefore if A then B" are tautologies) as errors in assertibility. But he also intends to analyze when we are epistemically justified in asserting a conditional and using it in our reasoning. Thus he has presented a direct rival to the semantic interpretation of intuitionistic logic given by Michael Dummett in *Elements of Intuitionism*. Rules for reasoning constitute a logic, and every propositional logic agrees with his analysis that (p. 71): "An indicative conditional is a certain kind of material conditional," as discussed above. But Jackson goes no further in presenting rules for reasoning than analyzing examples of conditionals with non-compound antecedent and consequent where the inference is a single step. Not only formally but practically we know that this is not enough, as in Example 7 above.[23]

Consider an example discussed earlier:

(‡) If winter comes early, I most likely will go to Costa Rica for Christmas.

I take this to be a conditional inference, to be analyzed in the same way as subjunctive conditionals. Jackson, however, distinguishes only between indicative and subjunctive, and only in virtue of their grammar. Since (‡) is in the indicative, his analysis should apply. So Jackson should say that (‡) is assertible if the probability of ("I will most likely go to Costa Rica" given that "Winter comes early") is high. But then we have the probability of an apparent probability claim, which he gives no guide to resolving. If we delete "most likely", his condition is just that the inference "Winter comes early therefore I will go to Costa Rica" is strong. Yet I argued above that (‡) is assertible (good) if that inference is only moderate and we can supply appropriate unstated premises. For each example Jackson considers, he supplies background conditions: small stories that place the conditional in context. He does not tell us how to proceed absent those.

---

[22] See footnote 20.

[23] If we try to analyze Example 7 according to Jackson's approach, we have no guide for whether we should consider the truth-value or the assertibility conditions of the antecedent, or neither.

# Explanations

Explanations are answers to questions. Verbal answers to a question why a claim is true can be evaluated as inferences that satisfy conditions peculiar to explanations. Some minimal conditions are typically taken as necessary, though not sufficient. Other conditions have been proposed, but they are either difficult to formulate clearly or have not been widely accepted. An important tool in evaluating inferential explanations is to recognize that the direction of inference of such an explanation is the reverse of that for an argument with the very same claims.

    Answers to a question about the function or goal of someone or something are teleological. They depend on clarity about the nature of functions and goals, and there is little agreement about criteria for those to be good beyond the necessity of avoiding the fallacy of assuming that because something occurs in nature it must have a purpose or goal.

Inferential explanations . . . . . . . . . . . . 128
Repairing explanations . . . . . . . . . . . . 132
Causal explanations . . . . . . . . . . . . . 133
Examples . . . . . . . . . . . . . . . . . . 135
Explanations and arguments . . . . . . . . . . 141
Explanations and predictions . . . . . . . . . 146
Comparing explanations . . . . . . . . . . . . 148
Inference to the best explanation . . . . . . . . 152
Causal explanations and causal laws . . . . . . 159
Explanations and theories . . . . . . . . . . . 166
Teleological explanations . . . . . . . . . . . 169
Conclusion . . . . . . . . . . . . . . . . . 172
    Appendix 1: Hempel's notion of a D-N explanation . . . 174
    Appendix 2: The likelihood principle . . . . . . . . 175

"Why does the sun rise in the east?" "How does electricity work?" "How come Spot gets a bath every week?" We give explanations as answers to lots of different kinds of questions.

## Inferential explanations

> **Inferential explanations** An answer to the question "Why is the claim E true?" which can be understood as "Because A, B, C, ... are true" is an *inferential explanation*. Often just the claims A, B, C ... are called the *explanation* of E. The claim E is sometimes called the *explanandum*, and A, B, C, ... the *explanans*.

Until the last section of this essay, I'll use "explanation" to mean inferential explanation and trust to context to make it clear when I'm using that word for the entire collection of claims or just the claims doing the explaining.

*Example 1* Zoe: Why is Spot limping?
　　Dick: Here, I see. It's because he's got a tiny thorn in his paw.
　　*Analysis* This is an inferential explanation: "Spot has a thorn in his paw" is meant to explain why "Spot is limping" is true. That is, the truth of "Spot is limping" is meant to follow from "Spot has a tiny thorn in his paw."

An inferential explanation is meant to be judged as an inference: The truth of the claim being explained is said to follow from the truth of the claims meant to do the explaining. What conditions are needed for such an inference to be a good explanation?

*The claim that's meant to be explained is plausible*
We can't explain what's dubious.

*Example 2* Why do most dogs like cats?
　　*Analysis* There can be no good explanation of this dubious claim.

*Example 3* Dick: Why is it that most people who call psychic hotlines are women?
　　Zoe: Wait a minute, what makes you think more women than men call psychic hotlines?
　　*Analysis* Dick has posed a **loaded question**: a question that presupposes that some dubious claim is true. Zoe has responded appropriately, asking for an argument to show that "More women than men call psychic hotlines" is true. Only if Dick can show that, would an explanation be called for.

*The explanation answers the right question*

*Example 4*

*Analysis* Flo thinks she has given a good explanation: Her answer makes it clear why the claim "There is only one piece of cake in the cupboard now" is true (assuming some other fairly obvious claims). But her mother won't accept it. Flo answered "Why is there only one piece of cake in the cupboard, instead of none?" but her mother meant, "Why is there only one piece of cake in the cupboard, instead of two?"

Questions are often ambiguous, and a good explanation to one reading of a question can often be a bad explanation to another. If the claim that is meant to be explained is ambiguous with respect to some intended interest of the person who posed the question, then that is a fault of the questioner; we should not be expected to guess correctly what's meant. But when it's obvious that a different reading of that sentence is meant, we are justified in classifying an explanation as bad, for it has explained the wrong claim.[1]

---

[1] Peter Lipton, "Contrastive explanation," and Bas C. van Fraassen, "The pragmatics of explanation," have developed analyses of explanations in terms of "contrastive classes." Van Fraassen says on p. 281: "Individual events are never explained, we only explain a particular event *qua* event of a certain kind."

As an example, "Why did Adam eat the apple?" can be understood as (a) "Why did *Adam* eat the apple?" or (b) "Why did Adam *eat* the apple?" or (c) "Why did Adam eat the *apple*?" The answer to (a) must contrast Adam with other possible eaters of the apple; the answer to (b) must contrast Adam's eating the apple with what else he might have done with it; the answer to (c) must contrast Adam's eating the apple with his eating something else. But that places the burden on the person explaining, when it is the responsibility of the person asking for the explanation to make clear that in, e.g., (a) the claim that is meant to be explained is "Why did Adam and not someone else eat the apple?"

*The claims doing the explaining are plausible*
In an inferential explanation the claims doing the explaining are supposed to make clear why the claim we are explaining is true. They can't do that if they aren't plausible.

*Example 5* The sky is blue because there are blue globules very high in the sky.
  *Analysis* This is a bad explanation because "There are blue globules very high in the sky" is not plausible.

  We don't require that in a good explanation the premises are true. An explanation is meant to explain to someone, to show to someone why a claim is true. So, as with arguments, it is meant to be judged by standards that are subjective, though generally intersubjective and perhaps based on objective standards that motivate our subjective ones. Whether an explanation is good is relative to what we think we know.

*The explanation is valid or strong*
The truth of the claim that's being explained is supposed to follow from the claims doing the explaining. So the relation between those claims should be valid or strong, like the relation between the premises and conclusion of a good argument.

*Example 6* Dogs lick their owners because dogs aren't cats.
  *Analysis* This is a bad explanation. The inference from "Dogs aren't cats" to "Dogs lick their owners" is neither valid nor strong, and there's no obvious way to repair it.

  As with arguments, *we allow that an explanation might need repair*. An explanation "E because of A, B, C, . . ." might require further claims to supplement A, B, C, . . . . But a good inferential explanation will have at least one claim among those that do the explaining that is less plausible than what's being explained. Otherwise it wouldn't explain, it would be a way to convince.

*The explanation is not circular*
We can't explain why a claim is true by just restating the claim in other words.

*Example 7* Zoe: Why can't you write today, Dick?
  Dick: Because I've got writer's block.

*Analysis* This is a bad explanation: "I've got writer's block" just means you can't write.

It is difficult to be precise about what we mean by an inference being circular. We cannot take the characterization:

> The claim that is meant to be explained is not equivalent to one premise or a conjunction of premises.

In general we're not able to discern such equivalences. We did not have that problem for arguments because we could subsume the restriction on non-circularity under the more general requirement that the premises be more plausible than the conclusion. But that does not apply here. At least, though, we can say that the explanation should not be "E because of D" where D is E itself or a simple rewriting of E.

Let's summarize these conditions.

---

*Necessary conditions for an inferential explanation to be good*
For the inferential explanation "E because of A, B, C, ..." to be good, all the following must hold:

- E is plausible.
- A, B, C, ... answer the right question.
- Each of A, B, C, ... is plausible, but at least one of them is not more plausible than E.
- The inference "A, B, C, ... therefore E" is valid or strong, possibly with respect to some other plausible claims.
- The explanation is not circular.

---

We sometimes say that an explanation is *right* or *correct* rather than "good," and *wrong* rather than "bad."[2]

With these conditions we can begin to analyze examples of explanations. In doing so we'll see other conditions that have been proposed for explanations to be good. Some of those criteria are quite difficult to make precise, and there is often disagreement about whether they are really necessary. There is certainly no agreement on what

---

[2] Some say that what I call a bad explanation is no explanation at all, or only a "potential explanation." However, the classification of some inferences as bad is continuous with classifications of other kinds of inferences (bad argument, false or unacceptable conditional, false causal claim).

constitute sufficient conditions for an explanation to be good. So when I make a judgment in what follows that an explanation is good after showing that it passes the necessary conditions we've adopted, that's only because I think it would seem good to most of us.

## Repairing explanations

Most explanations are not set out as inferences. Responding to a question, people typically leave out the conclusion, giving only the premises, just as some arguments have an unstated conclusion. When are we justified in interpreting what someone has said as an explanation? And when and how are we justified in repairing an apparently defective explanation? We make the same assumptions about people we used in argument analysis.

---

***The Principle of Rational Discussion***  We assume that the other person with whom we are deliberating or whose reasoning we are evaluating:
- Knows about the subject under discussion.
- Is able and willing to reason well.
- Is not lying.

---

We can then modify the Guide to Repairing Arguments by deleting the requirement that a premise that's added must be more plausible than the conclusion.

---

***The Guide to Repairing Explanations***  Given an (implicit) explanation that is apparently defective, we are justified in *adding* one or more premises or a conclusion if and only if both:
- The inference becomes valid or strong.
- The premise is plausible and would seem plausible to the other person.

If the inference is valid or strong, yet one of the original premises is implausible, we may *delete* that premise if the inference becomes no worse. In that case we say the premise is ***irrelevant***.

---

And as with arguments, there are clear conditions when an explanation is unrepairable.

*Unrepairable explanations* We cannot repair a (purported) explanation if any of the following hold:
- There is no explanation there.
- The explanation is so lacking in coherence that there's nothing obvious to add.
- A premise is implausible or several premises together are contradictory and cannot be deleted.
- The obvious premise to add would make the inference weak.
- Any obvious premise to make the inference strong or valid is implausible.
- The claim being explained is dubious.
- It answers the wrong question.

## Causal explanations

When an inferential explanation is given in terms of cause and effect, *if it's good causal reasoning and it answers the right question, then the explanation is good*; otherwise it's bad.

*Example 8* —Why did Dick wake up?
—Because Spot was barking.
*Analysis* This is a good explanation if Spot's barking really did cause Dick to wake up.

*Example 9* Dick recovered from his cold in one week because he took vitamin C.
*Analysis* This is a causal explanation, but not a good one: the purported cause does not clearly make a difference: almost everyone recovers from a cold in that time anyway.

*Example 10* Zoe: I wish I could help Wanda. What's the reason for her weight problem?
Dick: Gravity.
*Analysis* This is a causal explanation, but not a good one. The existence of gravity is a normal condition for Wanda weighing too much, not the or a cause.

*Example 11* Zoe: You say that this argument is bad. But why?
 Dr. E: It's bad because it's weak. For example, Sheila could have been a rabbit or a herring.
 *Analysis* Dr. E knows what he's talking about, and this is a good inferential explanation. But it's not a causal one. Explanations in terms of rules or criteria aren't causal.

*Example 12* The barometer is falling because a storm is coming.
 *Analysis* This is usually taken as a good explanation. The inference is valid or strong, and that can be established either via a generalization, based on lots of previous experiences, or via some relatively complicated meteorological claims. The latter establish something more: A storm is coming if and only if the barometer is falling. Compare then:

($\ddagger$)  A storm is coming because the barometer is falling.

David-Hillel Ruben says that Example 12 is a good explanation, but ($\ddagger$) is not an explanation. He suggests that we have to look for epistemological differences between these in order to see why.[3]

The only epistemological difference we need to invoke is that it is obviously true that the barometer is falling, while it is not as plausible that a storm is coming. So ($\ddagger$) is not a good explanation, though it is a good argument. If it's obvious that a storm is coming, then the difference between Example 12 and ($\ddagger$) could be explained by saying that Example 12 is a good causal explanation while ($\ddagger$) is not, since the barometer begins to fall only after a storm is coming.

*Example 13* Why is it that this pendulum has a period of 2.03 seconds? Because the pendulum is a simple pendulum; the length of the pendulum is 100 cm; the period $T$ of a simple pendulum is related to the length $L$ by the formula $T = 2\pi (L/g)^{1/2}$, where $g = 980$ cm/sec$^2$.
 *Analysis* Peter Achinstein takes this to be a good explanation.[4] He compares it to the same explanation with "This pendulum has a period of 2.03 seconds" and "The length of the pendulum is 100 cm" reversed, which he says is bad: "A pendulum has the period it has because of its length, but not vice versa." But the only reason I can

---

[3] *Explanation*, pp. 7–8.
[4] "Can there be a model of explanation?", p. 146. See also his "The pragmatic character of explanation."

see for that not being good is that we accept a correlation as providing an explanation only in the direction of cause to effect.[5]

## Examples

*Example 14*  Customer: Why did you call your coffee house *The Dog & Duck*?
  Owner: Why not?
  *Analysis*  Shifting the burden of proof is just as bad for explanations as for arguments.
  But aren't there times when "Why not?" is a good response? To assume otherwise is to assume that every true claim has a good explanation for why it is true, which is Leibniz's Principle of Sufficient Reason.[6]

*Example 15*  Suzy: Why did Dick just get up and leave the room like that in the middle of what Tom was saying?
  Zoe: Because he wanted to.
  *Analysis*  This is a bad explanation. Wanting to leave the room when Tom is talking is something unusual and requires further explanation. An explanation is *inadequate* if it leads to a further "Why?" Even if the claims doing the explaining are obviously true, they may not be enough.

---

[5] A similar example is discussed by Wesley Salmon in "Four decades of scientific explanation": Why is the shadow of the flagpole 30 feet here? Because the height of the flagpole is 18 feet and the sun is over there. Given a sufficient knowledge of trigonometry and light, plus a calculation of the angle of the sun above the horizon, the explanation is good. He says:

> Similarly, given the foregoing facts about the position of the sun and the length of the shadow, we can invoke the same law to deduce the height of the flagpole. Nevertheless, few people would be willing to concede that the height of the flagpole is explained by the length of its shadow. (p. 47)

One direction of the use of criteria is ruled out as an explanation because it is not causal. Bas C. Van Fraassen in "The pragmatics of explanation," Section 3.2, pp. 285–286, concocts a story to show that in some contexts the inference from the length of the shadow being 30 feet to the flagpole being 18 feet does constitute a good explanation. But in his story it isn't the length of the shadow that accounts for (is the cause of ) the height, but someone's wish that the shadow be that length.

[6] See Example 43 of "Reasoning about Cause and Effect" in this volume.

*Example 16*  Dick: Why won't Spot eat his dog food today?
Zoe: He just hasn't been hungry all day.
Dick: I don't think so—he just ate the doggy treat I gave him.
*Analysis*  Dick has shown that Zoe's explanation is not good by showing that the the claim doing the explaining is implausible.

*Example 17*  Zoe: Why did the lights just go out?
Dick: The transformer down the street must have blown again. I'll call the electric company.
Zoe: Don't bother—I can see the lights are still on where Flo lives, and the street lights are working.
*Analysis*  Zoe has used reducing to the absurd to show that Dick's explanation is wrong. If Dick's explanation were right, then from the same claim we could conclude, "The street lights and all the lights on the block are out." But that claim is false. So the claim doing the explaining is false.

*Example 18*  Dick: Why is your car sputtering like that?
Suzy: Because the battery is low.
Dick: C'mon. That's irrelevant.
*Analysis*  Suzy offers an explanation of "My car is sputtering." Dick's says that her claim is irrelevant, and it is: The inference from "The battery is low" to "Suzy's car is sputtering" is weak, with no obvious repair.

*Example 19*  Dick: Why did that turtle cross the road?
Lee: Because its leg muscles carried it.
*Analysis*  This is bad because it's the wrong kind of answer. We understand Dick to be asking for a behavioral explanation: claims about the motives or beliefs or feelings of a person or thing. What Lee gave is a claim about the physical make-up of the turtle, a physical explanation. It answers the wrong question.

On the other hand, "To get to the other side" would be an inadequate explanation. Since we don't know what the turtle's motives are, or indeed if it has any, it's unlikely that Lee or anyone else can give a good answer to Dick.

*Example 20*  Harry: I can't believe my uncle Ralph gave up his partnership in his law firm. He was making really big bucks. He sold his home and bought a cabin out near Big Tree Meadow, really primitive, and he says he's meditating. He was always such a responsible guy.

Zoe: How old is he?
Harry: 45.
Zoe: He's just going through mid-life crisis.

*Analysis* This is an inferential explanation, but we need more evidence for the claim that he's going through a mid-life crisis before we accept it as good—unless you define "mid-life crisis" as what Harry's uncle is doing, in which case the explanation is circular.

*Example 21* Zoe: My mom is always hot and irritable now.
Dick: Why's that?
Zoe: She's going through menopause.

*Analysis* This is a good causal explanation because there's good evidence that most women who go through menopause have those symptoms.

*Example 22* Psychiatrist: Where is Dr. E? It's time for his appointment.
Secretary: Don't you remember? He said he wouldn't be coming anymore.
Psychiatrist: He is resisting the understanding that I am bringing him.
Secretary: But he says it's because he can't afford your fee.
Psychiatrist: There, there's the proof that his unconscious is resisting, because I know he could borrow money from his rich uncle.

*Analysis* Psychiatrists often make their explanations immune to testing. If anything counts as resistance, if everything can be explained in terms of unconscious motives, there's no way to test. We might as well say that a patient won't come because gremlins are inhabiting his psyche, though that might be less effective in getting a patient to continue treatment and pay the bills. *If a claim explains everything, it explains nothing.* This is a bad causal explanation. *Untestable claims are the worst candidates for a good explanation.*

*Example 23* The slave boy knew Pythagoras' theorem because he could recollect it from a time before he had a human shape.[7]

*Analysis* Many people would say that this is a bad explanation because the claim doing the explaining cannot be "tested" (even theoretically) by observation. The explanation is not "scientific."

What we take to be a good explanation is relative to the metaphysics to which we subscribe.

---

[7] Compare Plato's *Meno*, 84–86.

*Example 24* Why did the water cool after it was placed over a flame for awhile and then removed?

> The water had initially been deprived of its glubbification through action of the flame (which generally has this effect); removed from the flame, it has become increasingly glubbified again and hence cooler, in accord with the principle that glubbification varies inversely with warmth.[8]

*Analysis*: Why is this explanation any worse than others in science that depend on highly abstract introduced technical terms?

What does "glubbify" mean? If it means just that things glubbify when they get cooler and de-glubbify when they get warmer, then the explanation is circular. Unless we have a definition of "glubbify" that is independent of this example and is supported by other evidence, the explanation is bad.

*Example 25*

Flo: Why did the window break?
Dick: Because the window is brittle
> *Analysis* What does "brittle" mean?[9]
> Easily broken or shattered.[10]
> Liable to break; fragile; friable.[11]

That the window broke because it's the kind of thing that breaks is a circular explanation, and hence bad. Dispositions to act in certain ways

---

[8] From *The Anatomy of Inquiry*, p. 127, by Israel Scheffler.
[9] The example (minus the cartoon) has been discussed at length by Gilbert Ryle, *The Concept of Mind*, p. 88, and Carl G. Hempel, "Aspects of scientific explanation," p. 457.
[10] *Webster's New World Dictionary, College Edition,* 1962.
[11] *The Shorter Oxford English Dictionary,* 1970.

are useless unless accompanied by some definition of the disposition and independent reasons for believing that the object has that disposition. As Wim J. van der Steen says,

> In ethology, concepts such as "drive" and "instinct" easily lead to circles. The argument that birds build nests because they have a nest-building instinct is no good if the instinct is merely defined as the disposition to build nests, the disposition being equated with the occurrence of nest-building under appropriate circumstances. If that is done the argument is a logical circle.[12]

A reviewer disagreed, saying "It's brittle" does not explain much, but it does locate the cause in the nature of the material, as opposed to, say, a defect in its manufacture, or its being already under stress from a twisted frame, or . . . . But this leaves Flo with at best an inadequate explanation, for she will want to know why glass is brittle.[13]

---

[12] *A Practical Philosophy for the Life Sciences,* p. 42.

[13] Robert Cummins, "Functional analysis," p. 758, says:

> Dispositions require explanation: if $x$ has $d$; then $x$ is subject to a regularity in behavior special to things having $d$, and such a fact needs to be explained.

M. F. Burnyeat, in "Virtues in action," p. 230, says as part of his argument that virtue is not a disposition:

> It is not enough to follow Ryle's account of dispositions, dividing them into two main types, tendencies (including traits of character like vanity) and capacities, and holding that to ascribe a tendency to something is to say that its state or behaviour is usually or often of a certain sort, or that it is so always or usually or often when certain particular conditions or kinds of conditions are fulfilled, while a capacity is constituted by its not being predictable that a subject will not [sic] behave in a certain way under certain conditions. Such things might be predictable or unpredictable on a variety of grounds, whereas a disposition must be some stable aspect of the subject which helps to explain (and is therefore not simply to be equated with) what is or is not to be expected from it. It can be explanatory, moreover, only where its issue is fixed for any recurrence of the conditions of its actualization, which may, of course, be various, with various outcomes, not all of them known. But if none is known with any definiteness, it will be equally vague what disposition is under discussion. Even the accredited dispositional terms of ordinary language, "brittle," "elastic," and the like, are affected to some extent by this sort of imprecision.

*140   Explanations*

*Example 26*

*Analysis*   An argument is meant to convince, but it can be good even if it doesn't convince. An explanation is meant to answer a question, but can it be good even if it isn't clear to someone that their question is answered? Can an explanation be good for one person but bad for another? Here's what Michael Scriven says:

> When we talk of "the explanation of sunspots," we rely on the fact that to get to the point where one understands the term "sunspot," but does not understand what produces them, is to acquire a fairly definite minimum body of knowledge. "The explanation" will be the one appropriate for a person with this knowledge, and this secondary sense of explanation is quite strong enough to survive certain types of failure of comprehension, i.e., a number of cases where "the explanation of X" is produced without the least understanding by those who hear it. In particular, the failure of two-year-old children to understand the explanation of sunspots is no ground for supposing the explanation to be unsatisfactory; it is "the explanation," not because it is understandable to everyone, but because it is understandable to the group that meets the conditions.[14]

The Principle of Rational Discussion applies to explanations. Dick's

---

[14] "Explanations, predictions, and laws," p. 205. On p. 207 he continues:
> Hence the notion of the proper context for giving or requesting an explanation, which presupposes the existence of a certain level of knowledge and understanding on the part of the audience or inquirer, *automatically entails* the possibility of a complete explanation being given. And it indicates exactly what can be meant by the phrase "*the* (complete) explanation." For levels of understanding and interest define areas of lack of understanding and interest, and the required explanation is the one which relates to these areas and not to those other areas related to the subject of the explanation but perhaps perfectly well understood or of no interest.

explanation appears bad to Zoe because she doesn't know enough about the subject. Whether the explanation is good is as objective or intersubjective as whether an argument is good.

But Zoe can reply that she understands light and water and oars as well as the next person, certainly well enough to ask the question. If Dick wants to maintain that he's given a good explanation, then he'll have to justify that it is a good explanation. That is, he'll have to continue explaining until he reaches a point where Zoe understands. We can call the original a good explanation, or only the entire collection of explanations a single good explanation.

In this and the last section we've seen the following additional conditions for an explanation to be good:

- A good causal explanation must be a good causal inference.

- If a behavioral explanation is expected and a physical one is offered, the explanation is not good, for it does not answer the right question. Similarly, if a physical explanation is expected and a behavioral one is offered, the explanation is not good.

- What counts as a good explanation is relative to the metaphysics to which we subscribe.

- Untestable claims do not explain well unless they are part of the assumptions we make in our metaphysics.

- Explanations in terms of dispositions are bad unless the disposition is itself explained.

An explanation that is inadequate and hence bad for one person may be a good explanation for another. Creating intersubjective, if not objective standards for what constitutes a good argument depends on judicious applications of the Principle of Rational Discussion.

## Explanations and arguments

*Example 27*  Dick: Ohhh. My head hurts.
   Zoe: You drank three cocktails before dinner, a bottle of wine with dinner, then a couple of glasses of brandy. Anyone who drinks that much is going to get a headache.

*Analysis* Zoe offers a good explanation of why Dick has a headache:

> Anyone who drinks that much is going to have a headache.
> Therefore (explains why), Dick has a headache.

Judged as an argument, however, this is bad, for it begs the question: it's a lot more obvious to Dick that he has a headache than that anyone who drinks that much is going to have a headache. The point of the explanation, though, is not to convince Dick or anyone else that the conclusion is true. The point is to show what the conclusion follows from. In a good explanation, the conclusion will be at least as plausible as any of the premises. *A good explanation is not a good argument.*[15]

*Example 28* Dick, Zoe, and Spot are out for a walk in the countryside. Spot runs off and returns after five minutes. Dick notices that Spot has blood around his muzzle. And they both really notice that Spot stinks like a skunk. Dick turns to Zoe and says, "Spot must have killed a skunk. Look at the blood on his muzzle. And he smells like a skunk."

Dick has made a good argument:

> Spot has blood on his muzzle. Spot smells like a skunk.
> Therefore, Spot killed a skunk.

Dick has left out some premises that he knows are as obvious to Zoe as to him:

> Spot isn't bleeding.
> Skunks aren't able to fight back very well.
> Dogs try to kill animals by biting them.
> Normally when Spot draws a lot of blood from an animal
>     that's smaller than him, he kills it.
> Only skunks give off a characteristic skunk-odor that
>     drenches whoever or whatever is near if they are attacked.

Zoe replies, "Oh, that explains why he's got blood on his muzzle and smells so bad." That is, she takes the same claims and views them as an explanation, a good explanation, relative to the same unstated premises:

---

[15] This illustrates (with a little rewriting) that many Aristotelian syllogisms that beg the question should be judged not as arguments but as explanations, attempts to explain or codify, not to convince.

Spot killed a skunk
*explains why* Spot has blood on his muzzle and smells like a skunk.

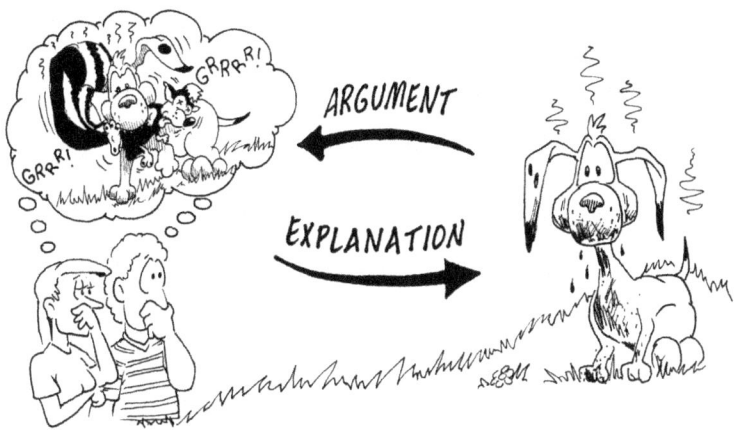

For an explanation "E because of A, B, C, ...", we can ask what evidence we have for A. Sometimes we can supply all the evidence we need just by reversing the inference. For Zoe's explanation to be good, "Spot killed a skunk" must be plausible, and it is because of the argument Dick gave—they needn't wait until they find the dead skunk.

---

***Explanations and associated arguments***  For an inferential explanation:

   E because of A, B, C, ...

the *associated argument* to establish A is:

   E, B, C, ... therefore A

An explanation is ***dependent*** if one of the premises is not plausible and the associated argument for that premise is not good. An explanation is ***independent*** if it is not dependent.

---

Note that there is an associated argument for each claim among the premises of an explanation.[16]

---

[16] Carl G. Hempel, "Aspects of scientific explanation," p. 372, calls independent explanations *self-evidencing*, but only if the inference is valid and the explanans contains a general law. Ernest Nagel, *The Structure of Science,* pp. 43–44, suggests that a necessary condition for an explanation to be good is that it be dependent. Michael Dummett, "The justification of deduction," p. 295,

144    Explanations

If an explanation is dependent, then it lacks evidence for at least one of its premises, which cannot be supplied by an associated argument.

*Example 29*   Spot chases cats because he sees them as something good to eat and because cats are smaller than him.
   *Analysis*   "Cats are smaller than Spot" is clearly true, but "Spot sees cats as something good to eat" is not. The associated argument for it is:

Spot chases cats and cats are smaller than Spot.
Therefore, Spot sees cats as something good to eat.

This is weak. Without more evidence for "Spot sees cats as something good to eat" we shouldn't accept the explanation, which is dependent.

Each premise of an independent explanation is plausible, either because of the associated argument for it or because of independent reasons, such as our knowing that most dogs bark or that Sheila is not a herring. Nonetheless, an independent explanation might be bad.

*Example 30*   Suzy: Why do classes last only fifty minutes instead of an hour?
   Maria: Because students need time to get from one class to another.
   *Analysis*   The premise is certainly plausible, so the explanation is independent. But the explanation is not good because it's not strong. Why not have classes that last forty-five minutes? What follows is only that there should be some time between classes.

*Example 31*   Suzy: Spot's licking you because he likes to taste the salt on your skin.
Dick: No. He's licking me because dogs are genetically programmed to behave that way.
Suzy: I don't believe you.
Dick: Sure. Tell him to come over to you very sternly.
Suzy: SPOT! Come here.
Dick: Now put your face next to him. See, he's licking you, and your skin isn't salty is it?
Suzy: No, you're right. I washed it just a little while ago.

---

among others, has also noted that the "epistemic direction" of explanations is the reverse of that of arguments. What's new is the observation that the direction of inference is reversed for the very same set of claims.

*Analysis* Dick offers an explanation of "Spot is licking you," namely, "Dogs are genetically programmed to behave that way." It is valid or strong with respect to some more or less obvious claims.

But why should we find that claim plausible? The associated argument "Spot is licking you, therefore dogs are genetically programmed to behave that way" is at best a weak generalization. This is a dependent explanation.

Dick does have other evidence for his explanation: Spot licks Dick. Every dog he's ever had has licked him. Every dog belonging to anyone he knows has licked their owner and friends of their owner, where "has licked" means "has the habit of usually licking their face when given the opportunity." Further, dogs are descended from wolves and share many of their characteristics. Wolf pups lick their mother to make the mother regurgitate so they can eat. This is not a learned habit, but appears to be genetically programmed. Wolves have been observed to lick other wolves superior to them.

We have an argument for "All dogs are genetically programmed to lick their superiors." The argument is a generalization, based on many observations plus some further claims that make the conclusion plausible. Dick believes the conclusion. Dick uses the conclusion of the generalization to explain why Spot is licking Suzy, and in doing so shows that Suzy's explanation is wrong.

*Example 32* This gas has temperature $83^{\circ}C$ because it has pressure 7 kg/cm$^2$ and volume 807 cm$^3$.

*Analysis* Here the claim "The gas has temperature $83^{\circ}C$," is explained by invoking both "The gas has pressure 7 kg/cm$^2$" and "The gas has volume 807 cm$^3$." Assuming both of these are true, this is a good explanation. The inference uses Boyle's Law as an unstated premise, along with a specification of what kind of gas this is and the constant of proportionality for that gas.

This is an example of a good explanation that is not cause and effect, since the premise does not become true before the conclusion. Each of the associated arguments is a good explanation, too. Any law of science that states a correlation can serve as a premise in an explanation, and that explanation, if good, will have an associated argument that is good. But that explanation need not be causal.

## Explanations and predictions

*Example 33*  Flo: Spot barks. And Wanda's dog Ralph barks. And Dr. E's dogs Anubis and Juney bark. So all dogs bark.

Barb: Yeah. Let's go over to Maple Street and see if all the dogs there bark, too.

*Analysis*  Flo, who's five, is generalizing. Her friend Barb wants to test the generalization.

Suppose that A, B, C, D are given as inductive evidence for a generalization G (some other plausible unstated premises may also be needed, but let's keep those in the background). Then we have that G explains A, B, C, D.

But if G is true, we can see that it follows that some other claims are true, instances of the generalization G, say L, M, N. If those are true, then G would explain them, too (Fido barks, Lady barks, Buddy barks). That is, G explains A, B, C, D and predicts L, M, N, *where the difference between the explanation and the prediction is that we don't know if the prediction is true*, not that the prediction is about the future.

Suppose we find that L, M, N are indeed true. Then the argument A, B, C, D + L, M, N, therefore G is a better one for G than we had before. At the very least it has more instances of the generalization as premises.

How can more instances of a generalization prove the generalization better? They can if (1) they are from different kinds of situations, that is, A, B, C, D + L, M, N cover a more representative sample of possible instances of G than do just A, B, C, D. Typically that's what happens: we deduce claims from G for situations we had not previously considered. And (2) because we had not previously considered the kind of instances L, M, N of the generalization G, we have some confidence that we haven't got G by manipulating the data, selecting situations that would establish just this hypothesis.

*A good way to test an hypothesis or generalization is to try to falsify it.* Trying to falsify a generalization means we are trying to come up with instances of the generalization to test that are as different as we can imagine from the ones we first used in deducing it. Trying to falsify is just a good way to ensure (1) and (2). We say that an experiment ***confirms***—to some extent—the (doubtful claims in the) explanation if it shows that a prediction is true.

*Explanations* 147

The story is much the same for claims that aren't generalizations. If A, B, C, D, E provide an argument for G, and G explains (has as consequences) L, M, N, O, P, we can check whether those are true. If they are, it is often the case that A, B, C, D, E + L, M, N, O, P provide a stronger proof of G. On the other hand, if one of those turns out to be false, then G is (most likely) false.[17] *Confirming an hypothesis-explanation is just strengthening the associated argument.*

*Example 34* The Atlanta Braves will win the pennant because they have the best pitching staff.

*Analysis* This sounds like an explanation. But it's not: "The Atlanta Braves will win the pennant" is not highly plausible. Rather, it is a prediction. If the prediction comes true, then it will be some evidence for a general claim like "The team with the best pitching staff always (usually?) wins the pennant."

*Example 35* Carl G. Hempel, in *Aspects of Scientific Explanation*, p. 365, tells the story:

> Consider the explanation offered by Torricelli for a fact that had intrigued his teacher Galileo; namely, that a lift pump drawing water from a well will not raise the water more than about 34 feet above the surface of a well. To account for this, Torricelli advanced the idea that the air above the water has weight and thus exerts pressure on the water in the well, forcing it up the pump barrel when the piston is raised, for there is no air inside to balance the outside pressure. On this assumption the water can rise only to the point where its pressure on the surface of the well equals the pressure of the outside air on that surface, and the latter will therefore equal that of a water column about 34 feet high.
>
> The explanatory force of this account hinges on the conception that the earth is surrounded by a "sea of air" that conforms to the basic laws governing the equilibrium of liquids in communicating vessels. And because Torricelli's explanation presupposed such general laws it yielded predictions concerning as yet unexamined phenomena. One of these was that if the water were replaced by

---

[17] Imre Lakatos in "A renaissance of empiricism" spells out his and Karl Popper's view that deduction (valid inferences via axioms in mathematics or science) is truth-preserving, taking truth from the axioms to their consequences, while inductive inferences are falsity-preserving: If the consequences are false, the premises are false. But only valid inferences are falsity-preserving; a strong inference only suggests strongly that conclusions are true or premises false.

mercury, whose specific gravity is about 14 times that of water, the air should counterbalance a column about 34/14 feet, or somewhat less than $2^{1}/_{2}$ feet, in length. This prediction was confirmed by Torricelli in the classic experiment that bears his name. In addition, the proposed explanation implies that at increasing altitudes above sea level, the length of the mercury column supported by air pressure should decrease because the weight of the counterbalancing air decreases. A careful test of this prediction was performed at the suggestion of Pascal only a few years after Torricelli had offered his explanation: Pascal's brother-in-law carried a mercury barometer (i.e., essentially a mercury column counterbalanced by the air pressure) to the top of the Puy-de-Dôme, measuring the length of the column at various elevations during the ascent and again during the descent; the readings were in splendid accord with the prediction.

*Analysis*  Torricelli offered an explanation, but the only evidence he had for the premise, which was a generalization, was the associated argument. So he made predictions: further instances of the generalization or of consequences of it. Those were shown to be true. The claim thus became more plausible because the associated argument for it was strengthened.

I am not, however, suggesting that every inferential explanation is a prediction of something we already know, or that every prediction is an explanation of something that isn't yet known to be true.[18]

# Comparing explanations

*Which is better?*
Given two explanations of the same claim, which is better? If one is right and the other wrong, the right one is better. If both are acceptable, we prefer the one that doesn't leave us asking a further "Why?".

We also prefer the simpler explanation.

---

[18] See Michael Scriven, "Explanations, predictions, and laws," especially Section 3.4, for a fuller critique of why explanations are not "essentially similar" to predictions. Israel Scheffler says in *The Anatomy of Inquiry*, p. 43:

> Prediction can neither be wholly assimilated to explanation as given in the deductive pattern nor, conversely, can explanation be understood as merely the provision of a potentially predictive base, a matter of showing that the problematic event *was to be expected* — as suggested by several writers. Explanation seems to require appeal to general principles, whether universal or statistical, serving to connect events in patterns.

**Simpler explanations**  One explanation of a claim is *simpler* than another explanation of that same claim if (i) its premises are more plausible, (ii) it is more clearly strong or valid (unstated premises are obvious and more plausible), and (iii) it has fewer steps.

Though this is not precise, it's certainly less vague than saying that one explanation is easier to understand.[19] All else then being equal, we prefer the stronger explanation.

*Example 36*  Zoe: How was your walk?
    Dick: Spot ran away again just before we got to the yard.
    Zoe: We better get him. Why does he run away just before you come home?
    Dick: It's just his age. He'll outgrow it. All dogs do.
*Analysis*  This sounded like a good explanation until Dick and Zoe found that Spot chased a cat up a telephone pole in the field behind their house. The explanation that Dick gave is not bad. Perhaps in a year or two when Spot is better trained, he won't run away even to chase a cat. But there's a better explanation: Spot ran away because he likes to chase cats and he saw a cat nearby to chase. It's better because it's stronger.

*Example 37*  Harry: Amazing. Zoe and Dick got into a fight and . . .
    Maria: Zoe's so mad she won't talk to him anymore.
    Tom: How do you know that?
    Suzy: She has ESP!
    Harry: C'mon. She must have heard it from someone else.
*Analysis*  Which is the better explanation? The second, because the claim doing the explaining is more plausible.

*Which is more general?*
Locating the claim we want explained within our general knowledge is one purpose of an explanation. So, it would seem, an explanation that is more general, putting the claim in a "bigger picture," is better. But what do we mean by "more general"?

---

[19] Norwood Russell Hanson in *Patterns of Discovery,* p. 95, says, "If that to which I refer when accounting for events needs more explaining than that to which you refer, then your explanation is better than mine." Philip Kitcher offers some development of these criteria in "Explanatory unification and causal structure," but only in the context of classical predicate logic.

Suppose that "$P_1, \ldots, P_r$ therefore E" is an explanation, a good one, even though, say, $P_7$ and $P_8$ are not obviously true. We would like to know why they are true. We want an explanation of them. So someone gives an explanation, say, "$Q_1, \ldots, Q_s$ therefore $P_7$ and $P_8$." $Q_1, \ldots, Q_s$ are less obviously true, but they are more general.

valid or very strong inferences ↓ | $Q_1, \ldots, Q_s$ therefore $P_7$ and $P_8$ and $P_1, \ldots, P_r$ therefore E | ↑ some evidence

---

***One explanation is more general than another***  $Q_1, \ldots, Q_s$ constitute a *more general explanation of* E than $P_1, \ldots, P_r$ if:

- Both are acceptable explanations of E.
- Possibly relative to some other claims that are plausible, any one of $P_1, \ldots, P_r$ that needs explaining is explained by $Q_1, \ldots, Q_s$, but at least one of $Q_1, \ldots, Q_s$ is not explained by $P_1, \ldots, P_r$.

---

This characterization does not exclude the possibility that the less general is actually contained as a part of the more general.

But a general explanation need not be better.

*Example 38* Nicholas Rescher compares the following pair of explanations:

> Why did this liquid turn solid? Because it is water and the temperature fell below 32° F, and all water solidifies when kept at a temperature below 32° F.
>
> Why did this liquid turn solid? Because that liquid was cooled to a very low temperature, and all liquids solidify when kept at a sufficiently low temperature.[20]

The second explanation is more general, but as Rescher says, the first is better.

> Not that generality isn't important—but its place is at another level, that of the explanation of laws. Thus consider . . . "Why did water

---

[20] *Scientific Explanation*, pp. 22–23.

solidify when kept at a temperature below 32° F?" Because water is a liquid and for all liquids there is a certain temperature below which they will solidify.[21]

*Analysis* The problem, though, is to give some clear idea of what is meant by a "law," as we'll see shortly.

*Example 39* Maria, Manuel, Lee, and Harry are camping in the woods. They're trying to make a fire, with little success:

Manuel: Why is it so hard to get a fire started?
Harry: Because $H_2O$ interferes with the uptake of oxygen combusting with the carbon in the wood.
Maria: It's 'cause the wood is wet.
Lee: Because it hasn't stopped raining and the wood can't dry out.

*Analysis* Harry, Maria, and Lee have given different explanations of why "It's hard to get a fire started" is true. Depending on what Manuel knows or wants to know, any one of these could be considered the best. If he doesn't know anything about chemistry and isn't interested, then Maria's is better than Harry's. If it's obvious that the wood is wet, then Lee's explanation may be better than Maria's. But if he's curious about how getting a fire started is related to the chemistry course he and Harry are taking, then Harry's answer is better than Lee's. Note that though Harry's explanation is the most "scientific," it need not be the best.

*Example 40* John Leslie in "The anthropic principle today" gives the following example:

Suppose you want to know why you have managed to catch a fish neither too big for your net, nor so small as to fall through it. Suppose the net is so ridiculously designed that only fish of an extremely limited range of sizes could be caught. Why did the net catch a particular fish at a particular time? A full explanation would include the entire history of this fish from birth onwards. But a useful partial explanation would be that there were very many fish in the lake, fish of very varied sizes, so that it was quite likely that an appropriately sized fish would sooner or later swim by.

"Yet doesn't the partial explanation possess some redundant features? Once we know the complete life-story of the fish which was caught, don't we see that the other fish were irrelevant?" True enough,

---
[21] Ibid.

the life-stories of the other fish might be separate, either totally or in all relevant respects, from that of the caught fish. If *to explain* a fish-catching could only mean *to give its causal history*, then the other fish might indeed fall outside the explanation, the relevant causal tale. Still, we ought to distrust the idea that this tale concerned *the one and only fish in the lake*. The existence of many fish would help make it believable that there was one of the right size. It would throw light on the affair. It would reduce or remove reasons for puzzlement. Now, "an explanation" can mean that sort of thing too. pp. 172–173

*Analysis* The first explanation, a "full" explanation, appears to assume a deterministic view of the universe: If we knew the full history of the fish we would see that it had to be caught. But a description of what happened is not an explanation, no matter how "full" it is. What is the difference between a history and a causal history? In any case, we do not have such "full" explanations, nor are we ever likely to have them, so the comparison is not apt: "Once we know the complete life-story of the fish which was caught" is fairy-tale talk. The partial explanation is good.

We have no agreement on criteria for what counts as a better explanation. What we seem to be able to do is compare pairs of explanations and generally agree on which is best.[22] But even if we could do that every single time, it still wouldn't be good evidence for there being an objective standard for "better explanation." We'll return to this issue of what constitutes a simpler or better explanation when we consider theories and explanations below.

## Inference to the best explanation

It can hardly be supposed that a false theory would explain, in so satisfactory a manner as does the theory of natural selection, the several large classes of facts above specified [the geographical

---

[22] Robin Horton in "Tradition and modernity revisited," says:

Most of the criteria so far devised for the critical assessment of efficacy in relation to the goals of explanation, prediction and control—e.g., simplicity, scope, degree of dependence on *ad hoc* assumptions, predictive power—are essentially relative rather than absolute in character. They are designed to tell us which of a pair of theoretical frameworks is, cognitively speaking, "better". They are not designed to tell us whether a single framework, considered in isolation, is "good" or "bad". p. 241

distribution of species, the existence of vestigial organs in animals, etc.]. It has recently been objected that this is an unsafe method of arguing; but it is a method used in judging of the common events of life, and has often been used by the greatest natural philosophers.
> Charles Darwin, *On the Origin of Species*, p. 476

If Darwin was right, why did scientists spend the next hundred years trying to confirm or disprove the hypotheses of natural selection? Only now do we believe that a somewhat revised version of Darwin's hypotheses are true.

Darwin is arguing backwards: From the truth of the conclusion(s), we can infer the truth of the premises. The direction of inference is wrong. Rather, the evidence for the claims doing the explaining comes from strengthening the associated argument. ***Darwin's mistake*** is to think that if we can easily deduce a lot of truths from a claim, then the claim must be true.[23]

*Example 41*  Tom: The AIDS epidemic was started by the CIA. They wanted to get rid of homosexuals and blacks, and they targeted those groups with their new disease. They started their testing in Africa in order to keep it hidden from people here. The government once again tried to destroy people they don't like.

*Analysis*  That the AIDS epidemic was started by the CIA would explain a lot. But that's no reason to believe it's true. The only evidence we have for it is the associated argument, which is weak. Every conspiracy theory depends on Darwin's mistake to make us believe it.

Gilbert Harman, however, thinks Darwin's method of arguing is right if the explanation is the best.

---

[23] Norwood Russell Hanson, *Patterns of Discovery*, p. 108, also thinks that a good explanation has to be true:

> If you accept the law of gravitation, the laws of Galileo and Kepler, the lunar motions and the tides will, as a matter of course, be systematically explained and cast into a universal mechanics.
>
> But why should I? The empirical truth of the law is not directly obvious, nor can what it asserts be easily grasped.
>
> Because if you accept it all these things will, as a matter of course, be systematically explained and cast into a universal mechanics. What could be a better reason?

Paul R. Thagard in "The best explanation: criteria for theory choice" presents further examples from the history of science where scientists reasoned this way.

154    *Explanations*

> In making this inference one infers, from the fact that a certain hypothesis would explain the evidence, to the truth of that hypothesis. In general, there will be several hypotheses which might explain the evidence, so one must be able to reject all such alternative hypotheses before one is warranted in making an inference. Thus one infers, from the premise that a given hypothesis would provide a "better" explanation for the evidence than would any other hypothesis, to the conclusion that the given hypothesis is true.
>
> There is, of course, a problem about how one is to judge that one hypothesis is sufficiently better than another hypothesis. Presumably such a judgment will be based on considerations such as which hypothesis is simpler, which is more plausible, which explains more, which is less *ad hoc,* and so forth. I do not wish to deny that there is a problem about explaining the exact nature of these considerations; I will not, however, say anything more about this problem.[24]

He calls this method of arguing ***inference to the best explanation***.[25]

For claims to explain satisfactorily, they have to be plausible—otherwise the explanation isn't good. Either the associated argument has to be good or we need independent evidence for the truth of them.

But if it's the best explanation, doesn't that count for it being true? We don't have accepted criteria for what counts as the best explanation, and in any case it's only the best explanation we've thought of so far. Scientists have high hopes for their hypotheses, and are motivated to investigate them if they appear to provide a better explanation than

---

[24] "The inference to the best explanation," p. 89.

[25] Richard Fumerton in "Inference to the best explanation," p. 208, points out that the explanation must be not only the best, but better than all others combined:

> It might be true of an explanation E1 that it has the best chance of being correct without it being probable that E1 is correct. If I have two tickets in the lottery and one hundred other people each have one ticket, I am the person who has the best chance of winning, but it would be completely irrational to conclude on that basis that I am likely to win. It is much more likely that one of the other people will win than that I will win. To conclude that a given explanation is actually likely to be correct one must hold that it is more likely that it is true than that the disjunction of all possible explanations is correct. And since on many models of explanation, the number of potential explanations satisfying the formal requirements of adequate explanation is unlimited this will be no small feat.

current theories. But the scientific community quickly corrects anyone who thinks that just making an hypothesis that explains a lot establishes that it's true.

---

***Fallacy of inference to the best explanation***  The *fallacy of inference to the best explanation* is to argue that because some claims constitute the best explanation we have, they're therefore true.

---

*Example 42*  Me: Why do I have such pain in my back? It doesn't feel like a muscle cramp or a pinched nerve.
Doctor: A kidney stone would explain the pain. Kidney stones give that kind of pain, and it's in the right place for that.
*Analysis*  When I went into the emergency room one night, the doctor gave me this explanation. It would have been a good one if he'd had good reason to believe "You have a kidney stone." But at that point the only reason he had was the associated argument, and that wasn't strong. Still, it was the best explanation he had.

So the doctor made predictions, reasoning by hypotheses: "A kidney stone would show up on an X-ray"; "You'd have an elevated white blood cell count"; "You would have blood in your urine." The next day he tested each of these and found them false. He then reasoned by reducing to the absurd that if the explanation were true, these would very likely be true; they are false; therefore, the explanation is very likely false.

Nothing else was found, so by process of elimination he concluded that I had a severe sprain or strain, for which exercise and education were the only remedy.

If the doctor had believed "You have a kidney stone" just because that was the best available explanation at the time, there would have been no point in doing tests. And then I would have undergone needless treatment or even surgery.

If we require for an inference to the best explanation that the explanation be not only better than all the others but also good, then we don't need inference to the best explanation. For an explanation to be good, the claims doing the explaining must be plausible.

Finding an explanation that's better than all others does not justify belief; it is only a good motive to investigate whether the claims doing the explaining are true.

*156   Explanations*

> *This is the best explanation we have.*
> *= This is a good hypothesis to investigate.*

Used this way, inference to the best explanation is often called *abduction*.[26]

*Example 43*   Flo's mother sees that a can of soda pop is gone from the refrigerator. She blames Flo for taking it. Flo denies it. Flo's mother says, "Well if you didn't take it, who did?"

*Analysis*   Even though inference to the best explanation is reasoning backwards, don't we do it all the time? Isn't it crazy to try to reform the whole world?

This example looks like inference to the best explanation: Flo taking the can of soda explains why it's gone. Flo's mother can't think of any other explanation. So Flo took it.

But the reasoning of Flo's mother can be better construed as:

Either Flo took the soda pop, or someone else did.
Flo likes soda pop and is likely to take some if she can.
(‡)  No one else had the opportunity.
Therefore, Flo took the soda pop.

All of the premises are plausible for Flo's mother. What she's doing is inviting Flo to give evidence that (‡) is false.

Many apparent uses of inference to the best explanation can be better understood as uses of a disjunctive syllogism. Seeing the reasoning in that light makes it much easier to evaluate it.

*Example 44*   Dick: There. See! The sign for the western men's wear store: "Real Men Don't Browse."

---

[26]  C. S. Peirce saw this clearly:

> The surprising fact, C, is observed.
> But if A were true, C would be a matter of course.
> Hence, there is reason to suspect that A is true.

> Deduction proves that something *must* be; Induction shows that something *actually is* operative; Abduction merely suggests that something *may be*.

The first quote is from *Collected Papers,* Vol. 5.189, reprinted in "Abduction and deduction," p. 151. The second is in *Collected Papers,* Vol. 5.171.

Bas C. van Fraassen analyzes inference to the best explanation in *Laws and Symmetry,* and he shows that even on the most generous interpretations it fails to be a good form of inference.

Zoe: Yeah. So.

Dick: It's true. We're genetically programmed that way. Long ago when humans were evolving, men were stronger and went out to hunt. The women gathered fruit and berries. When we're hunting we take the first thing we can get, maybe just a rabbit, even if we're looking for a mastodon, 'cause we don't know if there'll be another chance that day. When women went out to get berries there were lots of choices, so they'd pick the best and discard ones that weren't quite so good. That's why women like to shop and men just go in and buy the first thing that looks good and leave.

*Analysis* This explains why men don't browse and women do. But it doesn't explain it well. It's an example of Darwin's mistake, an evolutionary *just-so story*, like Rudyard Kipling's just-so story "How the Leopard Got Its Spots."

*Example 45* W.D. Hart in the Introduction to *The Philosophy of Mathematics*, p.6, reports on W.V.O. Quine's views:

> Sophisticated natural science as it comes is always to be formalized in an extension, in the logician's sense, of some mathematics, often number theory and analysis. Equations are obvious to anyone reading serious science. So, by abduction [inference to the best explanation], we are justified in believing true at least as much mathematics as we need for the best scientific explanations of what we observe. Since the truth of that much mathematics requires very abstract objects, Quine thereby began an empiricist justification for belief in the abstract objects required for mathematical truth.[27]

*Analysis* Inference to the best explanation is what stands behind many claims that there are numbers, and sets. Mathematics and science

---

[27] On pp. 5–6 he identifies Charles Peirce's notion of abduction (the previous footnote here) with Harman's inference to the best explanation. See Alan R. Baker, *Indispensability and the Existence of Mathematical Objects* for an historical summary of indispensability arguments.

A. A. Fraenkel, *Zehn Vorlesungen über die Grundlegung der Mengenlehre*, p. 61, makes an appeal that looks like inference to the best explanation within mathematics:

> The intuitive or logical self-evidence of the principles chosen as axioms [of set-theory] naturally plays a certain but not decisive role; some axioms receive their full weight rather from the self-evidence of the consequences which could not be derived without them.

"need" them, in the sense that they best explain why our mathematical and scientific theories are true. Certainly if they exist, they explain that. But, as in any use of inference to the best explanation, we must ask what other evidence we have to believe the claim "Numbers, as abstract objects, exist," since the inference from these scientific theories being "true" to numbers and sets existing is weak.[28]

Inference to the best explanation is no better in mathematics than in daily life. The difference, it seems, is that in mathematics there's no other evidence we can cite for "Numbers, as abstract objects, exist." Mathematics, for the platonist, is built on faith; and the necessity of numbers for mathematics—all numbers, natural, rational, real, in their abstract plentitude—is a guide, a sign towards that faith.[29]

Penelope Maddy also describes the "indispensability argument":

> We have good reason to believe our scientific theories, and mathematical entities are indispensable to those theories, so we have good reason to believe in mathematical entities. Mathematics is thus on an ontological par with natural science. Furthermore, the evidence that confirms scientific theories also confirms the required mathematics, so mathematics and science are on an epistemological par as well.[30]

Though the assumption that numbers exist as abstract objects may be compatible with our current mathematics, we have no need of it to do mathematics and apply mathematics to science.[31] Indeed, it's not clear that mathematicians use that assumption in any meaningful way.

---

[28] Mark Colyvan, in "Indispensability arguments in the philosophy of mathematics," says:

> Most scientific realists accept inference to the best explanation. Indeed, it might be said that inference to the best explanation is the cornerstone of scientific realism. But . . . inference to the best explanation may be seen as a kind of indispensability argument, so any realist who accepts the former while rejecting the latter is in a somewhat delicate position.

[29] Walter Carnielli and I suggest in *Computability*, p. 215 and Chapter 26, Section E, that faith cannot be banished from mathematics by adopting instead a formalist or constructivist view of numbers.

[30] "Indispensability and practice," p. 278. She criticizes the indispensability arguments on other grounds.

[31] See "On mathematics" in *Reasoning in Science and Mathematics* in this series. See also Solomon Feferman, "Why a little bit goes a long way: Logical foundations of scientifically applicable mathematics."

The indispensability argument has an unwelcome consequence: Abstract objects such as classes and numbers exist because the best science we now have requires them. But we are aware that our best science now may not be the best in the future. So these things might not exist—perhaps there are no numbers, but lots of other things, perhaps a huge different ontology.[32]

## Causal explanations and causal laws

Explanations are not meant to convince that the conclusion, the claim being explained, is true. They are meant to give us some insight into why it's true. But what is insight?

Consider what Galileo said when the laws for the motion of projectiles explained that if projectiles were fired from the same point with equal velocity but different elevations, the maximum range would be attained at an angle of 45°:

> From accounts given by gunners, I was already aware of the fact that in the use of cannon and mortars, the maximum range ... is obtained when the elevation is 45° . . .; but to understand why this happens far outweighs the mere information obtained by the testimony of others or even by repeated experiment.[33]

The insight, it is often said, is a law that connects the claim being explained to other experiences.[34]

But especially with causal explanations we often have no idea what general principle is needed, even though we accept the explanation as good.[35] Michael Scriven gives the following example:

> As you reach for the dictionary, your knee catches the edge of the table and thus turns over the ink-bottle, the contents of which proceed to run

---

[32] Maddy develops this idea in "Indispensability and practice," pp. 285–289.
[33] Galileo, *Dialogues Concerning Two New Sciences*, p. 265.
[34] See the discussion of causal laws in "Reasoning about cause and effect" in this volume for a discussion of why it is so hard to distinguish a claim that is a causal law from one that is an "accidental generalization."

Carl G. Hempel in "Aspects of scientific explanation," p. 337, says that we need to invoke laws in the explanans because of examples such as "Dick woke up because Dick woke up and Spot barked." The examples he invokes are causal, and the discussion of causal relevance in "Reasoning about Cause and Effect," pp. 53–54 in this volume, shows that no notion of law is needed to show that they are bad inferences.

[35] See the cartoon examples in "Reasoning about Cause and Effect" here.

over the table's edge and ruin the carpet. If you are subsequently asked to explain how the carpet was damaged you have a complete explanation. You did it, by knocking over the ink. The certainty of this explanation is primeval. It has absolutely nothing to do with your knowledge of the relevant laws of physics; a cave-man could supply the same account and be quite as certain of it. ... If you were asked to produce the role-justifying grounds for your explanation, what could you do? *You could not produce any true universal hypothesis* in which the antecedent was identifiably present (i.e., which avoids such terms as "knock hard enough"), and the consequent is the effect to be explained. ... The explanation has become not one whit more certain since the laws of elasticity and inertia were discovered.[36]

We can say that the claims doing the explaining must somehow link the claim being explained to other claims we know to be true, place it somehow within our general knowledge, show what it follows from. But at present only these vague desiderata are agreed upon for what we mean when we say that an explanation should give us some insight into why the claim being explained is true.[37]

---

[36] "Truisms as the grounds for historical explanations," p. 456.

[37] Michael Scriven, "Explanations, predictions, and laws," pp. 224-225, says:

What is a scientific explanation? It is a topically unified communication, the content of which imparts understanding of some scientific phenomenon. ... What is understanding? Understanding is, roughly, organized knowledge, i.e., knowledge of the relations between various facts and/or laws. ... It is for the most part a perfectly objective matter to test understanding, just as it is to test knowledge, and it is absurd to identify it with a subjective feeling, as have some critics of this kind of view. So long as we give examinations to our students, we think we can test understanding objectively.

John Stuart Mill in *A System of Logic*, III.XII.6, p. 337, says:

What is called explaining one law of nature by another, is but substituting one mystery for another; and does nothing to render the general course of nature other than mysterious: we can no more assign a *why* for the more extensive laws than for the partial ones. The explanation may substitute a mystery which has become familiar, and has grown to *seem* not mysterious, for one which is still strange. And this is the meaning of explanation, in common parlance. But the process with which we are here concerned often does the very contrary: it resolves a phenomenon with which we are familiar into one of which we previously knew little or nothing; as when the common fact of the fall

*Example 46* Why does Jupiter have a satellite? Because Jupiter has eight moons.

*Analysis*: We all agree this is a bad explanation. But why? Ernest Nagel uses this example to motivate the requirement that there must be a universal law among the claims doing the explaining in order for an explanation of an "individual event" to be good.[38] But this is a bad explanation simply because what is apparently wanted is a causal explanation, and this example is a bad causal inference: The premise does not become true before the conclusion does.

*Example 47* Why is the tide higher this week than last? Because the moon is in this part of its phase.

*Analysis* The inference is an adequate causal one. But Wesley Salmon says:

> Long before the time of Newton, mariners were fully aware of the correlation between the position and phase of the moon and the rising and falling of the tides. They had no knowledge of the causal connection between the moon and the tides, so they had no explanation for the rising and the falling of the tides, and they made no claim to any scientific explanation. To whatever extent they thought they had an explanation, it was probably that God in his goodness put the moon in the sky as a sign for the benefit of mariners. Nevertheless, given the strict law correlating the position and phase of the moon with the ebb and flow of the tides, it was obviously within their power to construct D-N [valid inferential] explanations of the

---

of heavy bodies was resolved into the tendency of all particles of matter toward one another. It must be kept constantly in view, therefore, that in science, those who speak of explaining any phenomenon mean (or should mean) pointing out not some more familiar, but merely some more general, phenomenon, of which it is a partial exemplification; or some laws of causation which produce it by their joint or successive action, and from which, therefore, its conditions may be determined deductively. Every such operation brings us a step closer towards answering the question which was stated in the previous chapter as comprehending the whole problem of the investigation of nature, viz.: what are the fewest assumptions, which being granted, the order of nature as it exists would be the result? What are the fewest general propositions from which all the uniformities existing in nature could be deduced?

[38] *The Structure of Science*, p. 30.

*162   Explanations*

behavior of the tides. It was not until Newton furnished the causal connection, however, that the tides could actually be explained.[39]

I think by "causal connection" Salmon means there was no causal law in the deduction of the height of the tide from the position of the moon. That is, Salmon's claim that the mariners had no causal explanation and hence no explanation for the ebb and flow of the tides amounts to saying that the formula relating phases of the moon to heights of tides, a generalization established by experience, is not a causal law.

Perhaps we should divide explanations and causal inferences into those that are supported by laws and those that are not. All Salmon and others have to do is give a characterization of what a law is in order to make that distinction more than an heuristic to improve our explanations.[40]

*Example 48*   This patient has paresis. It must be because he has syphilis.

*Analysis*   The only people who get paresis have syphilis first. But only a small percentage of people who get syphilis get paresis, so the explanation is weak. At best we have that syphilis can cause paresis, a cause-in-population claim.

Scriven, among others, believes this is a good explanation because there is no other known cause, regardless of how weak the inference is.[41] Hempel, on the other hand, argues that this is not a good explanation: It shows that we need a general law in the premises for the explanation to be good.[42] Salmon says that the premise gives an explanation but cannot yield a prediction, since from knowledge that

---

[39] "Four decades of scientific explanation," p. 47; the definition of a D-N explanation is given in Appendix 1. Salmon provides no evidence that ancient mariners did not take the inference to be a good causal explanation nor that they attributed the explanandum to God's goodness. They couldn't have thought of it as a scientific explanation, because the notion of scientific explanation dates from the twentieth century.

[40] The characterization has to be independent of any of our current formal logics because no formal logic we currently have allows for the formalization of many of the claims we intuitively accept as laws, for example, "Gold is heavier than water" or "Water is $H_2O$." Alternatively, this example can be seen as a case of a theory that is not based on a path of abstraction, as discussed in "On models and theories" in *Reasoning in Science and Mathematics*.

[41] "Explanation and prediction in evolutionary theory," p. 480.

[42] "Aspects of scientific explanation," pp. 369–370.

someone has syphilis we cannot predict that he or she will have paresis. In comparison, he says, the premise of the previous example yields predictions but no explanation.

*Example 49* Tom: Why didn't my number come up in the lottery?
Dick: Because there was only a 1 in 100,000,000 chance it would.
*later*
Suzy: It isn't fair. Why did Sheila's number come up on the lottery?
Zoe: Because someone's number had to.
*Analysis*: Wesley Salmon says:

> [Richard Jeffrey] maintained that when a stochastic mechanism—e.g., tossing of coins or genetic determination of inherited characteristics—produces a variety of outcomes, some more probable and others less probable, we understand those with small probabilities exactly as well as we do those that are highly probable. Our understanding results from an understanding of that mechanism and a recognition of the fact that it is stochastic.
>
> An explanation of a particular fact is an assemblage of facts *statistically relevant* to the fact-to-be-explained *regardless of the degree of probability* that results.[43]

Jeffrey and Salmon should classify both parts of this example as good explanations since the same "stochastic mechanism" governs both: We understand them because we understand the probabilities.

But both Tom and Suzy are asking for causal explanations, and by that standard both explanations are bad. A mathematical description of probabilities cannot be a cause, because that is timeless. Sometimes the best we can say is that it just happened.

But are these good non-causal explanations? If they are, then I don't understand what Jeffrey and Salmon mean by "explanation." I don't think I understand why Sheila won the lottery by being told she had a 1 in 100,000,000 chance of winning. That's no explanation at all; it's an admission that we don't have an explanation, that we don't understand the "mechanism" except in terms of relative frequencies.

A lot rides on what we take to be intuitively acceptable explanations. Jeffrey and Salmon disagree with me. I give necessary

---

[43] "Four decades of scientific explanation," pp. 62 and 67. The reference is to Jeffrey's "Statistical explanation vs. statistical inference." I take it that by "inductive arguments" Salmon means arguments that should be judged on the scale from strong to weak.

conditions for an explanation to be good. Jeffrey and Salmon exhibit examples such as these to show that those are not really necessary. We are at an impasse. We have different intuitions about what are good explanations. Perhaps we could agree to differ by saying that I have given necessary conditions for inferential explanations. Then Salmon and Jeffrey are giving an analysis of a broader or perhaps just different relation on claims, *chance explanations*.

*Example 50* A bomb is connected to a Geiger counter. If the reading on the counter is sufficiently high, the bomb will go off. A piece of radium is placed next to the Geiger counter. After three weeks the Geiger counter is taken away; the bomb never goes off.

*Analysis* If the bomb had gone off, then since the likelihood of enough radium atoms decaying in the period was so high, we could attribute putting the Geiger counter next to the radium as the cause of the explosion.[44] But here, even though there was an enormously high probability there would be enough radioactive decay to trigger the bomb, there wasn't enough decay. So putting the Geiger counter and bomb contraption next to the radium didn't cause it to go off.

In accord with the quotes in the last example, Salmon and Jeffrey should say that the following is a good explanation:

> Why didn't the bomb go off? Because there was a very low but not nonexistent possibility that not enough radium would decay.

It's a chance explanation.

We don't understand why a claim is true by saying that it was, according to our best physical theories, possible. The problem is that according to quantum physics we can't give any further kind of explanation. But either weak inferences are good explanations at all physical levels, or they are not good explanations at any level, since, as this example shows, what happens at the quantum level can have consequences in our daily lives.

Salmon notes that if relevant statistical probabilities count as an explanation, then we have to reject:

*Principle I*
If a set of circumstances of type C on one occasion explains the occurrence of an event of type E, then circumstances of the same type C cannot, on another occasion, explain the nonoccurrence of

---

[44] See Example 23 of "Reasoning about Cause and Effect" in this volume.

an event of type E (or the occurrence of an event of a type E´ that is incompatible with E).[45]

Stating the probabilities of picking any number explains why both Tom lost and Sheila won in the lottery. This seems a pretty strong reason to reject chance explanations as good. But Salmon says:

> What are the hazards in rejecting Principle I? We have often been warned—and rightly so—that science has no place for theological or metaphysical "theories" that explain whatever happens. Suppose someone is critically ill. If the person dies, the loved one explains it as owing to "God's will." If the person recovers, they explain it as owing to "God's will." Whatever happens is explained in terms of the will of the Almighty. However comforting such "explanations" might be, they are vacuous because there is no independent way of determining what God wills. To rule out "explanations" of this sort we do not need to appeal to Principle I; it is sufficient to insist that scientific explanations invoke *scientific* laws and facts. Scientific assertions, including those employed for purposes of explanation, should be supported by evidence. If we are dealing with what Hempel and Oppenheim called "potential explanation," then the laws or theories involved must be capable of independent support.[46]

To avoid the problem of giving up Principle I it seems that Salmon is willing to invoke science to explain science and/or require that all explanations be supported by other evidence. It is circular or vacuous to say that a scientific explanation must be scientific, or at least there is no clear standard for that. We do not have criteria for what counts as a scientific law or fact.[47] Someone invoking God's will may claim plenty of evidence for it, just not evidence that Salmon would accept as scientific. But a good way to show that it is not scientific is by invoking Principle I.

Often in life we are willing and feel compelled to say, "It's just coincidence." Yet as heirs to the Newtonian-mechanical view of the world, we feel we could say what the cause is if we had enough information; it's just our ignorance. But to claim that there is a good causal explanation even though we don't know it is a matter of faith.

---

[45] "Four decades of scientific explanation," p. 178.
[46] Ibid., p. 179.
[47] See the section on causal laws in "Reasoning about Cause and Effect" in this volume.

With this example, physicists, too, say "It's just coincidence." Some physicists go further: They say that there can't be any good causal explanation; it's not our ignorance. Other physicists, particularly Albert Einstein, disagree and argue that the limitations of this theory of the world show that it cannot be a complete theory.[48]

## Explanations and theories

Often we say that a theory explains something, for example, Newton's theory of motion explains the movement of the planets in the skies. But that's wrong. If someone were to ask why the planets move in the sky, and you were to state Newton's theory, you can be pretty sure the reaction would be "Huh?" A theory explains nothing. It is a deduction in a theory, an inference from some of the claims in the theory and claims about a specific situation according to the rules of inference of the theory, that explains.

Theories, though, are often invoked as part of the way in which explanations lead to insights and understanding. Here is what Michael Friedman says:

> This theory [the kinetic theory of gases] explains phenomena involving the behavior of gases, such as the fact that gases approximately obey the Boyle-Charles law, by reference to the behavior of molecules of which gases are composed. For example, we can deduce that any collection of molecules of the sort that gases are, which obeys the laws of mechanics will also obey the Boyle-Charles law. How does this make us understand the behavior of gases? I submit that if this were all the kinetic theory did we would have added nothing to our understanding. We would have simply replaced one brute fact with another. But this is not all the kinetic theory does—it also permits us to derive other phenomena involving the behavior of gases, such as the fact that they obey Graham's law of diffusion and (within certain limits) that they have specific heat capacities that they do have, from the laws of mechanics. The kinetic theory effects a significant *unification* in what we have to accept. Where we once had three independent brute facts—that gases approximately obey the Boyle-Charles law, that they obey Graham's law, and that they have the specific heat capacities they do—we now have only one—that molecules obey the laws of mechanics. Furthermore, the kinetic theory also allows us to integrate the behavior of

---

[48] See N. P. Landsman, "When champions meet: Rethinking the Bohr-Einstein debate."

gases with other phenomena, such as the motions of the planets and of falling bodies near the earth. This is because the laws of mechanics also permit us to derive both the fact that planets obey Kepler's laws and the fact that falling bodies obey Galileo's laws. From the fact that *all* bodies obey the laws of mechanics it follows that the planets behave as they do, falling bodies behave as they do, and gases behave as they do. Once again, we have reduced a multiplicity of unexplained, independent phenomena to one. I claim that this is the crucial property of scientific theories we are looking for; this is the essence of scientific explanation—a science increases our understanding of the world by reducing the total number of independent phenomena that we have to accept as ultimate or given. A world with fewer independent phenomena is, other things equal, more comprehensible than one with more.[49]

In Friedman's talk is some idea of laws and "brute facts," which we know is difficult to assess.[50] More apt is his disclaimer "within certain limits." Laws and theories aren't true; at best they are true enough, useful, applicable to certain situations.[51] Theories unify by allowing for innumerable explanations based on them, explanations whose premises are true (enough) as instantiations of the schematic claims that make up a theory.

*Example 51*  Why do the stars revolve in the heavens? Because the earth turns and revolves around the sun.

*Analysis*  The stars appearing to revolve in the heaven is not strong evidence for the earth turning. The explanation is dependent. The inference is strong or valid with respect to some claims that are obvious plus a general theory of physics, geometry, and astrophysics that would be plausible to an astronomer.

Why is this explanation better than "The earth is fixed and the stars move around the earth"? The latter is more plausible and accords better with our immediate experience. Moreover, Einstein has convinced astrophysicists that there is no preferred frame of reference for motion.

---

[49] "Explanation and scientific understanding," pp. 14–15. To be charitable, we should understand "phenomena" in the quote as "claims about phenomena." Theories do not allow us to deduce phenomena, for if they did we'd have to be very careful when deducing that there will be an explosion.

[50] See the section on causal laws in "Reasoning about Cause and Effect" in this volume.

[51] This analysis of theories is presented in "Models and Theories" in *Reasoning in Science and Mathematics* in this series.

The usual story is that the Copernican theory "explains more." Though it does not accord with our immediate experience (the sun moves, not the earth), by assuming enough about the movement of the planets and stars, many more of our observations of the skies can be explained than by Ptolemy's theory.

Ptolemy's theory can explain those observations, too. But Ptolemy's theory requires more and more adjustments that seem *ad hoc* in order to account for the observations, that is, to derive them from the theory. It is more a sense of the order, and "rightness," and simplicity of Copernicus' theory that recommends it to us over Ptolemy's.

Further, using Copernicus' theory we can make predictions about what we will observe in the skies, and those are usually more or less right, whereas even knowing that those predictions have turned out to be correct, it is extraordinarily difficult to derive them from Ptolemy's theory. That is, Copernicus' theory is recommended to us for its utility in making true predictions, which then serve as further evidence for the claims doing the explaining.

The difficulty is how to make these comments precise.[52]

We have no agreed-on criteria for what counts as a better explanation, though given two explanations we can generally make an intersubjective comparison. That problem is magnified for theories, for we have no clear criteria for what counts as a simpler theory. Over time we seem to agree that one theory is simpler than another, such as the Copernican analysis of the movement of the sun and stars compared to the Ptolemaic analysis. But to say that the one theory was simpler

---

[52] Henri Poincaré in *Foundations of Science*, p. 354 (*The Value of Science*, p. 141), wants to justify the choice of Copernicus' theory over Ptolemy's on purely objective grounds:

> In Ptolemy's system, the motions of the heavenly bodies cannot be explained by the action of central forces; celestial mechanics is impossible. The intimate relations that celestial mechanics reveals to us between all the celestial phenomena are true relations; to affirm the immobility of the earth would deny these relations, that would be to fool ourselves.
>
> The truth for which Galileo suffered remains, therefore, the truth, although it has not altogether the same meaning as for the vulgar, and its true meaning is more subtle, more profound and more rich.

when first proposed is an anachronism, for it is the use and familiarity of making deductions in the one theory compared to the other that generally convinces us.[53]

## Teleological explanations

One day while cleaning out the small pond in my backyard I asked myself, "Why is there a filter on this wet-dry vacuum?" The vacuum had a sponge-like filter, but the vacuum sucked up water a lot faster without it. I wondered if I could remove the filter.

I wanted to know the function of the filter. A causal explanation could be given starting with how someone once designed the vacuum with the filter, invoking what that person thought would be its function. But most of that explanation would be beside the point. I didn't want to know why it's true that there's a filter on the vacuum, even though the truth of that claim is assumed in the question. I wanted to know the function of the filter. Some explanations should answer not "Why is this true?" but "What does this do?" or "Why would he or she do that?"

---

**Teleological explanations**   A *teleological explanation* is one that invokes goals or functions, or uses claims that can come true only after the claim(s) doing the explaining.

---

*Example 52*   —Why is the missile going off in that direction?
—Because it wants to in order to hit that plane.
*Analysis*   It's a bad anthropomorphism to ascribe goals to a missile: People, not missiles, have goals. We should replace this teleological explanation with an inferential one: "The missile has been designed to go in the direction of the nearest source of heat comparable to the heat generated by a jet engine. The plane over there in that direction has a jet engine producing that kind of heat." Often a teleological explanation is offered when an inferential one should be used.

One problem with teleological explanations is that if an explanation uses claims that can be true only after the claim being explained becomes true, then it can't be causal (the cause has to precede the effect) and it would seem that the future is somehow affecting the past.

---

[53] See "Models and Theories" in *Reasoning in Science and Mathematics* in this series for a fuller discussion of the difficulty of comparing theories.

Many teleological explanations can be recast as inferential ones to avoid that problem.

*Example 53*   Zoe jogs every day because of her health.
   *Analysis*   We do not need to puzzle over how something in the future, her health, makes Zoe jog now. This can be recast as an inferential explanation: Zoe jogs because she believes that it will make her healthy.

*Example 54*   Spot barks in order to keep strangers away.
   *Analysis*   This is a good explanation in terms of one of the functions of Spot barking—if we agree that Spot has intentions. That becomes clearer if we recast it as an inferential explanation: Spot barks because he wants to keep strangers away.

*Example 55*   Suzy: Why is Spot crouching in front of that hole?
   Zoe: In order to catch a mouse.
   *Analysis*: But what if there is no mouse? How can we distinguish this from Spot crouching in front of the hole in order to catch a cat? This teleological explanation can be recast as an inferential one about Spot's habits or beliefs, as well as what Zoe knows about the hole. We can distinguish one of those motives for Spot from another only by knowing about Spot's behavior and how we attribute those motives to him.

A request for a teleological explanation assumes that the object has a function, or that the person or thing has a goal or motive. That's part of what's being assumed. But often there is simply no motive, no function, no goal, or at least none we can discern. The right response, as to a loaded question, is to ask why we should believe there is a function or motive.

*Example 56*   Dick (picking his nose): Why do humans get snot in their nose that dries up and has to be picked away? I can't understand what good it does.
   Zoe: What makes you think there's a purpose? Can't some things just be? Maybe it just developed along with everything else.

---

**Teleological fallacy**   The *teleological fallacy* is to assume that because something occurs in nature, it must have a purpose.

---

Another problem with teleological explanations is that we don't have criteria for what counts as a good one. That's because we don't have a clear idea how to judge what counts as the function of something. We can, however, give some necessary conditions for a teleological explanation E because of A, B, C, . . . to be good:
- E should be highly plausible.
- The explanation should answer the right question.
- It shouldn't be circular.
- It shouldn't ascribe motives, beliefs, or goals to something that doesn't have those.

*Example 57* Why does the blood circulate through the body?
(1) Because the heart pumps the blood through the arteries and veins.
(2) In order to bring oxygen to every part of the body tissues.
*Analysis* The first explanation is a good causal one, if it answers the right question. The second appears to be a good teleological one.

*Example 58* Why do we dream? Dreams serve as wish-fulfillments to prevent interruption of sleep, which is essential to good health.
*Analysis* According to Freudians, this is a good explanation. And it's teleological. There seems to be no way to construe it as inferential, at least not in accord with the rest of Freud's theory.

Opinion now divides: Either this is an example of a good explanation that is truly teleological, or this example shows that Freudian theories of the unconscious are no good because they yield only teleological explanations.

*Example 59* Why do humans have hearts? In order to circulate their blood.
*Analysis* This is a teleological explanation in terms of the function of the object. Robert Cummins says,

> An attempt to explain the presence of something by appeal to what it does—its function—is bound to leave unexplained why something else that does the same thing—a functional equivalent—isn't there instead. . . . To "explain" the presence of the heart in vertebrates by appeal to what it *does* is to "explain" its presence by appeal to factors that are causally irrelevant to its presence. . . .
> 
> To explain why it is there, why such a thing exists in the place (system, context) it does—this does require specifying factors that causally determine the appearance of that structure or process.[54]

172  Explanations

The explanation, then, is bad, because a causal explanation is called for. The inference from "Hearts circulate blood" to "Humans have hearts" is weak. Compare:

> What does the heart do?

"It circulates blood" is a perfectly good answer to this question, for which we would not be tempted to give an inferential explanation.[55]

Ernest Nagel disagrees. At least in biology, he says:

> It seems that when a function is ascribed to a constituent element in an organism, the content of the teleological statement is fully conveyed by another statement that is not explicitly teleological and that asserts a necessary (or possibly necessary and sufficient) condition for the occurrence of a certain trait or activity of the organism. In the light of this analysis, therefore, a teleological explanation in biology indicates the *consequences* for a given biological system of a constituent part or process; the equivalent nonteleological formulation of this explanation, on the other hand, states some of the *conditions* (sometimes, but not invariably, in physicochemical terms) under which the system persists in its characteristic organization and activities.
>
> Consider, for example, the teleological statement "The function of the leucocytes in human blood is to defend the body against foreign microorganisms." Now whatever may be evidence that warrants this statement, that evidence also confirms the nonteleological statement "Unless human blood contains a sufficient number of leucocytes, certain normal activities of the body are impaired," and conversely.[56]

Though of considerable interest for reasoning, the subject of functions is too large to pursue here.[57]

## Conclusion

Explanations are answers to questions. Verbal answers to a question why a claim is true can be evaluated as inferences that satisfy conditions peculiar to explanations. There are some minimal

---

[54] "Functional analysis," pp. 745–746.
[55] Cummins suggests, "What we can and do explain by appeal to what something does is the behavior of a containing system." Ibid. p. 748.
[56] "The structure of teleological explanations," p. 325.
[57] Larry Wright in "Functions" surveys the variety of uses of the notion of functions in explanations.

conditions that are typically taken as necessary, though not sufficient. Other conditions have been proposed, but they are either difficult to formulate clearly or have not been widely accepted. An important tool in evaluating inferential explanations is to recognize that the direction of inference of such an explanation is the reverse of that for an argument with the very same claims.

Answers to a question about the function or goal of someone or something are teleological. They depend on clarity about the nature of functions and goals, and there is little agreement about criteria for those to be good beyond the necessity of avoiding the fallacy of assuming that because something occurs in nature it must have a purpose or goal.

## Appendix 1  Hempel's notion of D-N explanation

Carl G. Hempel proposed dividing explanations into two types.[58] Those which are meant as valid inferences he calls *deductive-nomological* (D-N). The word "nomological" is meant to indicate that among the premises there are "general laws," universal claims of some sort.

The other kind of explanations are those that are meant as very strong but not valid, what he calls *probabilistic-statistical* explanations. The word "statistical" is meant to indicate that among the premises are claims that are just like general laws except that the quantifier is a percentage.

This division suffers from the same problems as the division of arguments into inductive and deductive discussed in "Induction and deduction" in *The Fundamentals of Argument Analysis* in this series. In particular, why should the intent of the person giving the explanation be invoked, rather than our judgment whether the inference should be evaluated one way as opposed to the other? Further, Hempel allows only statistical premises and not ones like "Almost all dogs bark," because he is concerned with "science."

## Appendix 2  The likelihood principle

The likelihood principle is meant to state conditions for preferring one hypothesis or explanation over others.[59] Given an observation $O$, hypotheses $H_1$, $H_2$, and a probability function $P$:

*Likelihood principle*
$O$ favors $H_1$ over $H_2$ if and only if $P(O/H_1) > P(O/H_2)$.

This is reasoning backwards, a variant on inference to the best explanation. As Eliott Sober says:

> Both [inferences] involve "abduction," which is to say that in both cases data are interpreted by appeal to the Likelihood Principle.[60]

We are interested in how much the observation supports the hypotheses, and for that we should consider the strength of arguments from the observation to the hypotheses, not from the hypotheses to the observation.

One colleague who uses statistics in his work disagreed. As an example, he said, suppose one hypothesis $H_1$ about causes of breast cancer predicts that 10% of a particular population will have breast cancer, while another, $H_2$,

---

[58] See, for example, Carl G. Hempel, "Aspects of scientific explanation" or, for a short exposition by him, "Explanation in science and history."

[59] It was originally proposed by Ian Hacking, *Logic of Statistical Evidence*, pp. 54ff. The simplified version presented here is given by Elliott Sober, "Epistemology for empiricists," p. 47.

[60] "Epistemology for empiricists," p. 56. See footnote 26 above.

predicts that only 4% will have breast cancer. The observation $O$ is made that 9.87% of that population have breast cancer. Then by the likelihood principle, the evidence favors the first hypothesis. This is good reasoning, he said.

But we have a better reason for prefering the first hypothesis: The argument from the observation to that hypothesis is stronger than the argument from the observation to the second hypothesis, $P(H_1/O) > P(H_2/O)$ (roughly, if $H_1$ then $O$, if $H_2$ then not-$O$, but $O$, therefore $H_1$).

Though this application of the principle of likelihood can be justified in terms of the associated arguments, not all uses can. Consider:

A   All dogs bark.
B   Between 50% and 100% of all dogs bark.
O   All dogs I have observed bark.

The likelihood principle says we should favor $A$ over $B$. But the argument "$O$ therefore $B$" is stronger and hence better than "$O$ therefore $A$," and so we should prefer $B$ over $A$.

Another colleague says that this example does not overthrow the likelihood principle. It just shows that the following is false:

($\ddagger$)   $O$ favors $A$ over $B$   iff
    $O$ favors $A$ more strongly than $O$ favors $B$.

The left-hand side, he says, is a primitive notion, not reducible to the right-hand side. Arguments are meant to convince, lead to belief that a claim is true, while the relation of "favors more strongly than" is not meant to lead to any belief.[61]

But what is this three-place relation? It is not enough to say that science should be understood as giving us only reason to prefer one claim, based on the evidence, over another, rather than as leading to belief. That "prefers" means "gives more reason to believe," which, while perhaps not fully "reason to believe," is justified in exactly the same way, through the use of arguments.[62] The likelihood principle stands in need of justification, and cannot be taken as primitive.[63]

---

[61] As Ian Hacking says in *Logic of Statistical Evidence*, p. 28:

   If one hypothesis is better supported than another, it would usually be, I believe, right to call it the more reasonable. But of course it need not be reasonable positively to believe the supported hypothesis, nor the most reasonable one. Nor need it be reasonable to act as if one knew the best supported hypothesis were true.

[62] See *The Fundamentals of Argument Analysis* in this series.

[63] Neither Hacking, op. cit., nor Sober, op. cit., nor Richard M. Royall in his advocacy of the likelihood principle in *Statistical Evidence*, provide that kind of justification.

# Bibliography

Page references are to the most recent publication cited unless noted otherwise.

ACHINSTEIN, Peter
   1993a  Can there be a model of explanation?
          In RUBEN, 1993, pp. 136–159.
   1993b  The pragmatic character of explanation
          In RUBEN, 1993, pp. 326–358.

ANSCOMBE, G. E. M.
   1975  Causality and determination
        In SOSA, 1975, pp. 63–81.
        Reprinted in SOSA and TOOLEY, pp. 88–104.

ARISTOTLE
   1928  *The Works of Aristotle*
        ed. W. D. Ross, Oxford University Press. This includes:

| | |
|---|---|
| *Analytica Priora* | Trans. by W. J. Jenkinson. |
| *De Sophisticis Elenchis* | Trans. by W. A. Pickard–Cambridge. |
| *Physica* | Trans. by R. P. Hardie and R. K. Gaye. |

BAKER, Alan R.
   1999  *Indispensability and the Existence of Mathematical Objects*
        Ph.D. Thesis, Princeton University.

BLANSHARD, Brand
   1964  *The Nature of Thought*
        2 volumes. Humanities Press.

BORN, Max
   1949  *Natural Philosophy of Cause and Chance*
        Clarendon Press, Oxford.

BOYD, Richard
   1985  Observations, explanatory power, and simplicity: toward a non-Humean account
        In *Observation, Experiment, and Hypothesis in Modern Physical Science*, ed. P. Achinstein and O. Hannaway, The MIT Press, pp. 47–94. Reprinted in *The Philosophy of Science*, eds. R. Boyd, P. Gasper, and J.D. Trout, MIT Press, 1991, pp. 349–377.

BURKS, Arthur W.
   1977  *Chance, Cause, Reason*
        University of Chicago Press.

BURNYEAT, M. F.
  1980 Virtues in action
    In *The Philosophy of Socrates*, ed. Gregory Vlastos, University of Notre Dame Press, pp. 209–234.
CARNIELLI, Walter A. *See* EPSTEIN and CARNIELLI.
CHISHOLM, Roderick M.
  1946 Contrary-to-fact conditionals
    *Mind*, vol. 55, pp. 289–307.
    Reprinted in *Readings in Philosophical Analysis*, eds. H. Feigl and W. Sellars, Appleton-Century-Crofts, 1949.
  1955 Law statements and counterfactual inference
    *Analysis*, vol. 15, pp. 97–105. Reprinted in SOSA, 1975, pp. 147–155.
COHEN, Morris and Ernest NAGEL
  1934 *An Introduction to Logic and Scientific Method*
    Harcourt, Brace and Company.
COLYVAN, Mark
  1999 Indispensability arguments in the philosophy of mathematics
    *The Stanford Encyclopedia of Philosophy* (Fall 1999), http://plato.stanford.edu .
CUMMINS, Robert
  1975 Functional analysis
    *The Journal of Philosophy*, vol. LXXII, pp. 741–765.
DARWIN, Charles
  1872 *On the Origin of Species*
    Sixth edition. Reprinted, Collier, 1962.
DAVIDSON, Donald
  1967 Causal relations
    *Journal of Philosophy*, vol. 64, pp. 691–703.
    Reprinted in DAVIDSON, 1980.
    Reprinted in SOSA and TOOLEY, 1993, pp. 75–87.
  1969 The individuation of events
    In *Essays in Honor of Carl G. Hempel*, ed. Nicholas Rescher, D. Reidel.
    Reprinted in DAVIDSON 1980.
  1980 *Essays on Actions and Events*
    Clarendon (Oxford University Press).
DUCASSE, C. J.
  1926 On the nature and the observability of the causal relation
    *Journal of Philosophy*, vol. 23, pp. 57–68.
    Reprinted in SOSA and TOOLEY, 1993, pp. 125–136.

DUMMETT, Michael
  1964  Bringing about the past
          *The Philosophical Review,* vol. 73, pp. 338–359.
          Reprinted in DUMMETT, 1978, pp. 333–350.
          Page references are to the earlier publication.
  1973  The justification of deduction
          *Proceedings of the British Academy,* vol. LIX, 1975, pp. 201–232.
          Reprinted in DUMMETT, 1978, pp. 290–318.
  1978  *Truth and Other Enigmas*
          Harvard University Press.

EDWARDS, Paul, ed.
  1967  *The Encyclopedia of Philosophy*
          Macmillan & The Free Press.

EINSTEIN, Albert
  1915  *Relativity: The Special and General Theory*
          Fifteenth edition, 1961, Crown Publishers, Inc.

EINSTEIN, A., B. PODOLSKY, and N. ROSEN
  1935  Can quantum-mechanical description of physical reality be considered complete?
          *Physical Review,* vol. 47, pp. 777–780.

ELLINGTON, James W.
  1977  Introduction to *Prolegomena to Any Future Metaphysics* by Immanuel Kant
          Hackett Publishing Company.

EPSTEIN, Richard L.
  1990  *Propositional Logics*
          Kluwer. 2nd edition, Oxford University Press, 1995.
          2nd edition with corrections, Wadsworth, 2000.
  1994  *Predicate Logic*
          Oxford University Press. Reprinted, Wadsworth, 2000.
  1998  *Critical Thinking*
          3rd edition with Carolyn Kernberger, Wadsworth, 2005.
  2001  *Five Ways of Saying "Therefore"*
          Wadsworth.
  2006  *Classical Mathematical Logic*
          Princeton University Press.
  2010  *The Internal Structure of Predicates and Names with an Analysis of Reasoning about Process*
          Typescript at www.AdvancedReasoningForum.org .
  2011a *Reasoning in Sciences and Mathematics*
          Advanced Reasoning Forum.

2011b  *Prescriptive Reasoning*
       Advanced Reasoning Forum.
201?   *Reasoning and Formal Logic*
       To appear, Advanced Reasoning Forum.
201?   *The Fundamentals of Argument Analysis*
       To appear, Advanced Reasoning Forum.

EPSTEIN, Richard L. and Walter A. CARNIELLI
1989   *Computability: Computable Functions, Logic, and the Foundations of Mathematics*
       Wadsworth & Brooks/Cole.
       3rd edition, Advanced Reasoning Forum, 2008.

EWING, A. C.
1934   *Idealism: A Critical Survey*
       3rd edition, 1974, Methuen & Co. Ltd. and Barnes & Noble.

FARRELL, Daniel M.
1988   Recent work on emotions
       *Analyse & Kritik*, vol. 10, pp. 71–102.

FEFERMAN, Solomon
1993   Why a little bit goes a long way: Logical foundations of scientifically applicable mathematics
       *PSA*, vol. 2, (Philosophy of Science Association), pp. 442–455.
       Reprinted in Feferman, *In the Light of Logic*, Oxford Univ. Press. 1998, pp. 284–298.

FRAENKEL, A. A.
1927   *Zehn Vorlesungen über die Grundlegung der Mengenlehre*
       B. G. Teubner. The translation is from LAKATOS, 1967.

FRIEDMAN, Michael
1974   Explanation and scientific understanding
       *Journal of Philosophy*, vol. 71, pp. 5–19.

FUMERTON, Richard
1992   Inference to the best explanation
       In *A Companion to Epistemology*, eds. J. Dancy and E. Sosa, Blackwell, pp. 207–209.

GALILEI, Galileo
1949   *Dialogues Concerning Two New Sciences*
       Translated by H. Crew and A. de Salvio, Northwestern Univ. Press.

GOLDIE, Peter
2002   Emotions, feelings and intentionality
       *Phenomenology and the Cognitive Sciences*, vol. 1, pp. 235–254.

GOODMAN, Nelson
   1947   The problem of counterfactual conditionals
*Journal of Philosophy,* vol. 44.
Reprinted in *Semantics and the Philosophy of Language,*
ed. Leonard Linsky, Univ. of Illinois Press, 1952, pp. 231–246.

HACKING, Ian
   1965   *Logic of Statistical Inference*
Cambridge University Press.

HANSON, Norwood Russell
   1958   *Patterns of Discovery*
Cambridge University Press.

HARMAN, Gilbert
   1965   The inference to the best explanation
*Philosophical Review,* vol. 74, pp. 88–95.

HART, H.L.A. and Tony HONORÉ
   1985   *Causation in the Law*
2nd edition, Oxford University Press.

HART, W. D. , ed.
   1996   *The Philosophy of Mathematics*
Oxford University Press.

HEMPEL, Carl G.
   1962   Explanation in science and history
In *Frontiers of Science and Philosophy,* ed. R. G. Colodny,
published jointly by Allen & Unwin and University of Pittsburgh
Press. Also in RUBEN, 1993.
   1965c  *Aspects of Scientific Explanation*
The Free Press. This includes:
        Chapter 9. General laws in history.
          A revised version of an article of the same name in
          *The Journal of Philosophy,* vol. 39, 1942, pp. 35–48.
        Chapter 12. Aspects of scientific explanation.

HOLLIS, Martin and Stephen LUKES, eds.
   1982   *Rationality and Relativism*
The MIT Press.

HONORÉ, Tony. *See* HART and HONORÉ.

HORTON, Robin
   1982   Tradition and modernity revisited
In HOLLIS and LUKES, 1982, pp. 201–260.

HUME, David
   1739   *A Treatise of Human Nature*
Edition 1888, ed. L. A. Selby-Bigge, Oxford University Press.

JACKSON, Frank
　1987　*Conditionals*
　　　Basil Blackwell.
KANT, Immanuel.　*See* ELLINGTON.
KIM, Jaegwon
　1971　Causes and events: Mackie on causation
　　　*Journal of Philosophy*, vol. 68, pp. 426–441.
　　　Reprinted in SOSA and TOOLEY, 1993, pp. 60–74.
　1987　Explanatory realism, causal realism, and explanatory exclusion
　　　*Midwest Studies in Philosophy*, vol. 12, pp. 225–239.
　　　Reprinted RUBEN, 1993.
KITCHER, Philip
　1989　Explanatory unification and the causal structure of the world
　　　In *Scientific Explanation*, Kitcher and Wesley C. Salmon, eds.,
　　　Minnesota Studies in the Philosophy of Science Vol. XIII,
　　　University of Minnesota Press.
KVART, Igal
　1986　*A Theory of Counterfactuals*
　　　Hackett Publishing Company.
　1997　Cause and some positive causal impact
　　　*Philosophical Perspectives*, vol. 11, Blackwell Publishers,
　　　pp. 401–432.
LAKATOS, Imre
　1967　A renaissance of empiricism
　　　In *Problems in the Philosophy of Mathematics*, ed. I. Lakatos,
　　　North-Holland.
LANDSMAN, N. P.
　2006　When champions meet: Rethinking the Bohr-Einstein debate
　　　*Studies in History and Philosophy of Modern Physics*, vol. 37,
　　　pp. 212–242.
LE POIDEVIN, Robin and Murray MACBEATH
　1993　*The Philosophy of Time*
　　　Oxford University Press.
LEIBNIZ, Gottfried Wilhelm
　1875–90　*Die Philosophische Schriften von G.W. Leibniz*, I–VII
　　　ed. C. I. Gerhardt, Berlin. Reprinted by Hildesheim, 1965.
　1981　*New Essays on Human Understanding*
　　　Cambridge University Press.
LESLIE, John
　1997　The anthropic principle today
　　　In *Final Causality and Human Affairs*, ed. Richard F. Hassing,
　　　The Catholic University of America Press.

LEWIS, David
   1973  *Counterfactuals*
          Harvard University Press.
   1976  Probabilities of conditionals and conditional probabilities
          Philosophical Review, vol. 85, pp. 297–315.
   1986a Events, pp. 241–269.
          In Lewis, *Philosophical Papers, Vol. II*, Oxford University Press.
   1986b A subjectivist's guide to objective chance, pp. 83–113.
          In Lewis, *Philosophical Papers, Vol. II*, Oxford University Press.

LIPTON, Peter
   1990  Contrastive explanation
          In *Explanation and its Limits*, ed. D. Knowles, Cambridge University Press. Reprinted in RUBEN, 1993.
   1991  *Inference to the Best Explanation*
          Routledge.

LUKES, Stephen. *See* HOLLIS and LUKES.

LUTZ, Antoine, Lawrence L. GREISCHER, Nancy B. RAWLINGS, Matthieu RICARD and Richard DAVIDSON
   2004  Long term meditators self-induce high-amplitude gamma synchrony during mental practice
          *Proceedings of the National Academy of Sciences of the USA*, vol. 101, no. 46. Accessed at http://www.pnas.org/cgi/content/full/101/46/16369.

MACBEATH, Murray. *See* LE POIDEVIN and MACBEATH.

MACKIE, J. L.
   1967  Mill's methods of induction
          In EDWARDS, 1967, vol. 5, pp. 324–332.
   1973  *Truth, Probability and Paradox*
          Oxford University Press.
   1974  *The Cement of the Universe*
          Oxford University Press. With corrections and additions, 1980.

MADDY, Penelope
   1992  Indispensability and practice
          *Journal of Philosophy*, vol. 59, pp. 275–289.

MATES, Benson
   1986  *The Philosophy of Leibniz: Metaphysics and Language*
          Oxford University Press.

MILL, John Stuart
   1874  *A System of Logic, ratiocinative and inductive, being a connected view of the principles of evidence and the methods of scientific investigation*
          8th edition, Harper & Brothers Publishers (New York).

MONTAGUE, Richard
  1960  On the nature of certain philosophical entities
        *Monist,* vol. 53, pp. 159–194.
        Reprinted in *Formal Philosophy: Selected Papers of Richard Montague,* ed. R. Thomason, Yale Univ. Press, 1974.
MORAVCSIK, Julius
  1974  Aristotle on adequate explanations
        *Synthese,* vol. 28, pp. 3–17.
NAGEL, Ernest
  1949  In defense of logic without metaphysics
        *The Philosophical Review,* vol. LVIII.
        Reprinted in *Logic without Metaphysics,* The Free Press, 1956, pp. 93–102.
  1961  *The Structure of Science*
        Harcourt, Brace & World.
        Reprinted by Hackett Publishing Company, 1979.
  1984  The structure of teleological explanations
        In SOBER, 1984, pp. 319–346, from NAGEL, 1961, pp. 398–428.
        *See also* COHEN and NAGEL.
PEIRCE, Charles
  1960  *Collected Papers,* vol. 5 and vol. 6
        eds. Charles Hartshorne and Paul Weiss, The Belknap Press, Harvard.
PLATO
  1961  *The Collected Dialogues of Plato*
        eds. Edith Hamilton and Huntington Cairns, Princeton University Press. This includes:
            *Phaedo*  Trans. by Hugh Tredennick.
            *Meno*    Trans. by W. K. C. Guthrie.
POINCARÉ, Henri
  1921  *The Foundations of Science*
        Translated by G. B. Halstead, The Science Press, New York.
  1958  *The Value of Science*
        Translated by G. B. Halstead, Dover.
POPPER, Karl
  1972  *The Logic of Scientific Discovery*
        Hutchison of London. A translation and revision of *Logik der Forschung,* Vienna, 1935. Oxford University Press, 1971.
PROSSER, William L.
  1955  *Handbook of the Law of Torts*
        2nd edition. West Publishing Co.

REISENZEIN, Rainer
   1995   On appraisals as causes of emotions
*Psychological Inquiry*, vol. 6, pp. 233–237.

RESCHER, Nicholas
   1970   *Scientific Explanation*
The Free Press.
   1996   *Process Metaphysics*
State University of New York Press.

ROSENFELD, L.
   1967   Niels Bohr in the thirties. Consolidation and extension of the conception of complementarity
In S. Rozental, ed., *Niels Bohr: His Life and Work as Seen By His Friends and Colleagues*, North-Holland, pp. 114–136.

ROYALL, Richard
   1997   *Statistical Evidence*
Chapman & Hall.

RUBEN, David-Hillel, ed.
   1993   *Explanation*
Oxford University Press.

RUSSELL, Bertrand
   1912   On the notion of cause, with applications to the free-will problem
*Proceedings of the Aristotelian Society*, 1912–1913. Also in *Mysticism and Logic*, Longmans, Green and Co., 1925, pp. 180–208.

RYLE, Gilbert
   1949   *The Concept of Mind*
Hutchinson.

SALMON, Wesley C.
   1984   *Scientific Explanation and the Causal Structure of the World*
Princeton University Press.
   1989   Four decades of scientific explanation
In *Scientific Explanation*, Minnesota Studies in the Philosophy of Science Vol. XIII, University of Minnesota Press, ed. P. Kitcher and W. Salmon.
Reprinted as a separate book by the University of Minnesota Press, 1990. Page references are to the earlier publication.
   1993   Causality: production and propagation
In SOSA and TOOLEY, 1993, pp. 154–171.

SCHEFFLER, Israel
   1963   *The Anatomy of Inquiry*
Bobbs-Merrill. Reprinted by Hackett Publishing Company, 1981.

SCOTT, Dana
  1971  On engendering an illusion of understanding
        *The Journal of Philosophy,* vol. 68, pp. 787–807.
SCRIVEN, Michael
  1959a Truisms as the grounds for historical explanations
        In *Theories of History,* ed. P. Gardiner, Free Press, pp. 443–475.
  1959b Explanation and prediction in evolutionary theory
        *Science,* vol. 130, pp. 477–482.
  1962  Explanations, predictions, and laws
        In *Minnesota Studies in the Philosophy of Science,* vol. III, eds. H. Feigl and G. Maxwell, University of Minnesota Press, pp. 170–230.
SEARLE, John R.
  1999  I married a computer
        *New York Review of Books,* Vol. XLVI, no. 6 (April 8), pp. 34–38.
SKLAR, Lawrence
  1978  Up and down, left and right, past and future
        *Noûs,* vol. 15, pp. 115–129.
        Reprinted with emendations in Sklar, *Philosophy and Spacetime Physics,* University of California Press, 1985.
        Reprinted in LE POIDEVIN and MACBEATH, 1993, pp. 99–116.
SKYRMS, Brian
  1986  *Choice & Chance*
        Third edition. Wadsworth.
SOBER, Elliott
  1993  Epistemology for empiricists
        *Midwest Studies in Philosophy,* XVIII, pp. 39–61.
  1984  *Conceptual Issues in Evolutionary Biology*
        The MIT Press.
SOSA, Ernest, ed.
  1975  *Causation and Conditionals*
        Oxford University Press.
SOSA, Ernest and Michael TOOLEY, eds.
  1993  *Causation*
        Oxford University Press.
STALNAKER, Robert
  1968  A theory of conditionals
        In *Studies in Logical Theory,* ed. Nicholas Rescher,
        *APQ Monograph No. 2,*Basil Blackwell.
STRAWSON, P. F.
  1966  *The Bounds of Sense*
        Methuen.

TAYLOR, Richard
   1966   The metaphysics of causation
           Originally part of *Action and Purpose,* Humanities Press.
           Reprinted in SOSA, 1975, pp. 39–43.
   1967   Causation
           In EDWARDS, 1967, vol. 2, pp. 56–66.

THAGARD, Paul R.
   1978   The best explanation: criteria for theory choice
           *The Journal of Philosophy,* vol. 75, pp. 76–92.

THODE, E. Wayne
   1968   The indefensible use of the hypothetical case to determine cause in fact
           *Texas Law Review,* vol. 46, pp. 423–435.

TOOLEY, Michael
   1987   *Causation: A Realist Approach*
           Oxford University Press.
           *See also* SOSA and TOOLEY.

UNWIN, Nicholas
   1996   The individuation of events
           *Mind,* col. 105, pp. 315–330.

VAN DER STEEN, Wim J.
   1993   *A Practical Philosophy for the Life Sciences*
           State University of New York Press.

VAN FRAASSEN, Bas C.
   1989   *Laws and Symmetry*
           Clarendon Press.
   1993   The pragmatics of explanation
           In RUBEN, 1993, pp. 275–309.

VLASTOS, Gregory
   1969   Reasons and causes in the *Phaedo*
           *Philosophical Review,* 78, pp. 291–325.
           Reprinted in *Platonic Studies,* by Gregory Vlastos, Princeton University Press, 1981, pp. 76–110.

VON WRIGHT, Georg Henrik
   1973   On the logic and epistemology of the causal relation
           From *Logic, Methodology and Philosophy of Science IV,* eds. P. Suppes et al. North-Holland.
           Reprinted in SOSA and TOOLEY, 1993, pp. 105–124.

WAISMANN, Friedrich
  1945  Verifiability
        *Proceedings of the Aristotelian Society* Supp. Vol. 19, 1945, pp. 119–150.
        Reprinted in *The Theory of Meaning*, ed. G.H.R. Parkinson, Oxford University Press, 1968, pp. 33–60.
WALLACE, William A.
  1972  *Causality and Scientific Explanation*
        In 2 volumes (2nd volume, 1974), University of Michigan Press.
WALTERS, R. S.
  1967  Contrary-to-fact conditional
        In EDWARDS, 1967, vol. 2, pp. 212–216
WHITE, Morton G.
  1965  *Foundations of Historical Knowledge*
        Harper & Row. Harper Torchbooks edition, 1969.
WHITEHEAD, Alfred North
  1925  *Science and the Modern World*
        Macmillan Co. Reprinted by The Free Press, 1976.
WRIGHT, Larry
  1973  Functions
        *Philosophical Review*, vol. 82, pp. 70–86.
        Reprinted in SOBER, 1984, pp. 347–368.

# Index

*italics* indicate a definition or quote
n indicates a footnote

abduction, *156*–157, 175
abstract objects, 22–24
accidental generalization, 77, 80.
  *See also* law, causal
Achinstein, Peter, 134
*aitia*, 82
analogy, *11*
Anscombe, G.E.M., 43n, 44n
antecedent, *102*
arguing backwards, *12*, 153
argument, *4*
  associated, *143*
  causal inference vs., 21
  explanation vs., 141–145
  good, *5*
  necessary conditions to be good, *7*
Aristotelian syllogisms, 142n
Aristotle, *81*–82, 83, 90
associated argument, *143*

backwards reasoning, *12*, 153
Baker, Alan R., 157n
"becausal" claims, 17, 99, 100
begging the question, *6*
best explanation. *See* inference to the best explanation.
biased sample, *10*
Blanshard, Bland, *90*–91
Born, Max, *43*, 91
Boyd, Richard, 42n, 92
burden of proof in causal analyses, 33

Burks, Arthur, 91
Burnyeat, M.F., *139n*
butterfly causation, 56

Carnielli, Walter, 158n
cartoon examples, 30–37
causal chain, 55, 57
causal claim, *14*
  general, *14, 20*
  hypothetical, *64*–65, 114n, 120
  particular, *14*
  precedence over general?,
    75–77
  *See also* causal inference; cause
causal explanation, 92, 130–131,
  133–135, 137, 139, 145, 152
  causal laws and, 159–166
causal inference, *20, 27*, 90–93
  arguments vs., 21
  burden of proof, 33
  explanations as, 92
  good, *20*
  necessary conditions to be good,
    minimal, *18*–19
  usual, *28*
  strong can be good?, 21, 37,
    42–44, 47, 78
  sufficient conditions to be
    good?, 29–30
causal law. *See* law, causal
causal power, 29n, 40–42, 60n,
  89, 98
Causation, Law of, 66–67
cause
  a —, *50*, 57
  close in time and space to
    effect, 24–25, 28, 54–55
  common, 26
  conditional isn't, 23

cause (continued)
  constant conjunction and, 38.
    See also law, causal;
    generalization
  definition isn't, 57–58
  described with a claim, 14
  efficient cause, 82–84
  final cause, 82, 83, 85
  finding a, 67–70
  formal cause, 82, 83
  functional relations instead, 41–42
  happens, 22
  human agent as, 55–56
  in population, 70–75, 71
  intervening, 27, 54
  law needed for. See law, causal
  legal system and. See legal
    system and causes
  makes a difference, 19. See
    also sine qua non condition
  manipulability criterion, 60
  material cause, 82
  minimal notion of, 18–21, 20
  need not be something active,
    15–16
  normal conditions for cause and
    effect, 18–19
    distinguished from cause, 47–50
    See also law, causal; generalization; strong causal inference
    can be good?
  overdetermination, 50–51
  precedes effect, 22. See also
    teleological explanation
  proximate, 55–57
  quantified claim isn't, 23
  reason vs., 20, 23, 46, 65, 81–85, 90
  simultaneous with effect?, 22,
    57–62, 65
  sine qua non condition, 46–47,
    51, 57. See also cause, makes a
    difference

static conditions, 15–16
subjective claim describes, 52
time and space, 21–24
tracing backwards, 32, 37,
  50–51, 56
usual notion of, 27–28
See also things as causes
chain, causal, 55, 57
chance explanation, 164–166
chance, objective. See objective
  chance
change, 15–16, 61
Chisholm, Roderick, 121
circular explanation, 92, 130–131,
  139
claim, 1
  dubious, 5
  equivalent, 1
  necessary, 107
  plausible, 5–6
  possible, 107
classical abstraction, 105
classical modal logic, 108
classical propositional logic, 105
Cohen, Morris R., 76
Colyvan, Mark, 158n
common cause, 26, 59, 61
comparing explanations, 148–152
Comte, Auguste, 41n, 42n
conclusion, 2
  follows from premises, 4
conditional, 102
  conditions to be true, 105
  counterfactual, 112n, 121–125
  generalized, 103
  material, 104
  subjunctive, 112n, 121,
    124–125
conditional inference, 111
  good, 114
conditionalization of an inference,
  106

confirming an explanation,
    146–148
consequent, *102*
controlled cause-to-effect experiment,
    72
convincing, 4–5
Copernican theory, 167–168
correlation-causation fallacy, *75*
counterfactuals, 64, 77, 112n,
    121–125
Cummins, Robert, *139, 171*

Darwin, Charles, *152–153*
Darwin's mistake, *153*
Davidson, Donald, 85–86
Deduction Theorem, *106*
definition isn't cause, 57–58
dependent explanation,
    *143*–145
Descartes, 84
determinism, 67, 152
dispositions, 138–139
D-N explanation, 161, 174
drawing the line fallacy, *2*
dubious claim, *5*
Ducasse, C.J., 60n, *75–76*
Dummett, Michael, 22n, 125,
    143n

effect. *See* cause
Einstein, Albert, 62n, 167
Ellington, James, *59*
entailment, *107*
entropy, 22*n*
equivalent claims, *1*
events, 16, 17, 21, 24, 48, 51,
    53–54, 62, 64, 65–66, 69,
    75–77, 84, 85–89
Ewing, A.C., *90*
experiment (controlled,
    uncontrolled), 72
explanandum, *128*

explanans, *128*
explanation,
    Aristotle's four kinds of causes
        as, 82n
    best. *See* inference to the best
        explanation
    causal, 92, 130–131, 133–135,
        137, 139, 145, 152. *See also*
        law, causal
    chance, *164*–166
    confirming an, *146*–148
    dependent, *143*–145
    independent, *143*–145
    inferential, *128*
    argument vs., 141–145
    associated argument of, *143*
    best. *See* inference to the
        best explanation
    circular, 130–131, 139
    comparing explanations,
        148–152
    D-N, 161, 174
    explains everything, 137,
        164–166
    more general, 149–152,
        *150*, 166–168
    necessary conditions to be
        good, 131
    predictions and, 146–148,
        162–163, 168
    repairing, 130
    simpler, *149*
    teleological, *169*–172
    necessary conditions to be
        good, 170
    teleological fallacy, *170*
    theory as an —, 166–169

false cause, 90
falsifying a prediction, 146–147.
    *See also* testable claim
Farrell, Daniel M., 95n

fear as cause, 16, 51–52
Feferman, Solomon, 158n
final cause, 82, 83, 85
finding a cause, 67–70
foreseeable consequence, 54
formal cause, 82–83
Fraenkel, A.A., *158n–159*
Freud, S., *171*
Friedman, Michael, *166–167*
Fummerton, Richard, *154n*
functional relations vs. cause, 41–42

Galileo, 147, *159*, 167
Geiger counter example, 43–44, 165–166
general causal claim, *14*, *20*
generalization, *10*
  accidental, *77*, 80
  necessary conditions to be good, *11*
  needed for causal analysis?, 32, 33, 34, 37, 38–42, 48, 54
  statistical, *10*
  *See also* law, causal; general causal claim
generalized conditional, *103*
God, 79, 84, 85, 161, 165
Golden, Anne L. *74–75*
Goldie, Peter, 97n
Goodman, Nelson, *121*
Guide to Repairing Arguments, *9*, 113, 132
Guide to Repairing Explanations, 132

Hacking, Ian, 174n, *175n*
Hanson, Norwood Russell, *48–49*, *79–80*, *93*, 149n, *153n*
Harman, Gilbert, 153–*154*, 157n
Hart, H.L.A. and Tony Honoré, *16*, *28n*, *47–48*, *51*, 53n, *53–54*, *55–56*, *56–57*, *62–63*, *76–77*

Hart, W.D., *157*
heart example, 171–172
Hempel, Carl G., *92*, 138n, 143n, *147*, 159n, 162, 174
Honoré, Tony. *See* Hart, H.L.A.
Horton, Robin, *152n*
human agent in causal analyses, 55–56
Hume, David, *29n*, 76–77, 79, 89, 92
*hypotheses non fingo*, 83
hypotheses, reasoning by, 111–112
hypothetical causal claim, *64–65*

implication, strict, *107*–110
inadequate explanation, *135*
independent explanation, *143*–145
indicator words, 2
indispensability argument, 157–159
inductive evidence, *10*
inertia of objects, 52
inference, 2
  conditional, *111*
  conditionalization of, *106*
  strong, *3*
  valid, *3*
  weak, *3*
inference to the best explanation, 152–159, *154*
  fallacy of, *155*
inferential explanation. *See* explanation, inferential
intervening cause, *27*, 54
intuitionism, 105, 118, 119, 125
irrelevant premise, *9, 132*

Jackson, Frank, 103n, 124–125
Jeffrey, Richard, 163–164
just-so story, *157*

Kant, 22n
Kelly, Jennifer L., *74–75*
Kim, Jaegwon, 29n, *87*
Kitcher, Philip, 149n
Kvart, Igal, 50n, 53n, *87n*, 89n, 121, *124*

Lakatos, Imre, 147n
Landsman, N.P., 61n, 166n
law, causal, 23n, 41–42, 46n, 75–80, 159–166, 167. *See also* general causal claim; generalization; law, scientific; legal system and causes; normal conditions
Law of Causation, 66–67
law, scientific, 43, 44n, 52, 71–72, 108, 110, 165, 166. *See also* law, causal
Le Poidevin, Robin, 22n
legal system and causes, 28n, 51, 53–54, 55–57, 62–63
Leibniz, *66, 82–83*, 109n
Lewis, David, *87*, 89n, *103n*, 123
likelihood principle, *174*–175
Lipton, Peter, 129n
loaded question, *128*
Lutz, Antoine, et al., 96n

Macbeath, M., 22n
Mackie, J.L., *51n*, 52n, *86–87n, 89*, 121n
Maddy, Penelope, *158*–159n
manipulability criterion for cause, 60
material cause, 82
Mates, Benson, *83, 84, 88, 109n*
Method of Induction, Mill's, 69
Mill, John Stuart, *47, 49n, 66–67, 69, 160n*
modal logic, 107–110
  classical, 108

modal words, *107*
*modus ponens, 104*
*modus tollens, 112n*, 115–116
Montague, Richard, *86*
Moravcsik, Julius, 75b
Mundt, Diane J., *74–75*

Nagel, Ernest, 29n, 143n, 161, *172*
names, non-referring, 65
necessary claim, *107*
necessity, 108–110
Newton, Isaac, 41n, 71, 83–84, 161–162, 166
normal conditions for a conditional inference, 114
normal conditions for cause and effect, *18–19*
  distinguished from cause, 47–50
  *See also* law, causal; generalization; strong causal inference can be good?
nothing, 65
numbers exist, 157–159

objective chance, 21, 44n, 89–90
objects. *See* things as causes; things as objects of emotions
objects, inertia of, 52
open sentence, *103*
overdetermination, causal, 50–51

particular causal claim, *14*
Peirce, Charles S., *156n*, 157n
pencil example, 58–59
Philo of Megara, 104
physics, 40–41, 44, 52, 61, 79
Plato, 137n. *See also* Socrates
plausible claim, *5*–6
Podolsky, B., 62
Poincaré, Henri, *168n*

Popper, Karl, 92, 147n
population, *10*
population, cause in, 70–75, *71*
possible claim, *107*
possible world, 77, 109–110, 122–123
*post hoc ergo propter hoc*, 36, 38, 75
power, causal 29n, 40–42, 60n, 89, 98
predicate logic, 23, 64, 87, 91, 149n
predictions and explanations, 146–148, 162–163, 168
premise, *2*
  irrelevant, *9*
Principle I, *164*–166
Principle of Rational Discussion, 8, *132*, 140–141
Principle of Sufficient Reason, 66, 135
processes, 17–18, 23, 24n, 60
proposition, 1
Prosser, William, *57n*
proximate cause, 54–57
Ptolemaic theory, 167–168

Quine, W.V.O., 157

Rational Discussion, Principle of, 8, *132*, 140–141
reason vs. cause, 20, 23, 46, 65, 81–85, 90
reasoning backwards, *12*
*reductio ad absurdum*, 114, 116, 120, 136
regularist, 76–79
relevance, causal, *53*, 57
relevant premise, *9*
repairing arguments, 8–9, 113
representative sample, *10*
Rescher, Nicholas, 23–24n, 78–79, 121–122, *150*–*151*
Resenzein, Rainer, 99n

Rosen, N., 62n
Rosenfeld, L., 62n
Royall, Richard, 175n
Ruben, David-Hillel, 134
Russell, Bertrand, 23n, 61, 84
Ryle, Gilbert, 138n, 139n

Salmon, Wesley, *24n, 135n, 161*–*162, 163*–*165*
sample, *10*
  biased, *10*
  randomly chosen, *11*
  representative, *10*
Scheffler, Israel, 138n, *148n*
scientific law. *See* law, scientific
Scott, Dana, 110n
Scriven, Michael, *48, 67, 140*, 148n, *159*–*160*, 162
Searle, John R., *42n*, 96
self-referential paradoxes, 110n
simultaneous cause and effect?, 22, 57–62, 65
*sine qua non* condition, *46*–47, 51, 57
Sklar, Lawrence, 22n
Skyrms, Brian, *89*–*90*
Sober, Eliot, *174*, 175n
Socrates, *45*–46, 57, *81*
space. *See* time and space in causal reasoning
Stalnaker, Robert, *122*–*123*
static condition as cause, 15–16, 60–61, 87. *See also* change
Strawson, P.F., *60n*
strict implication, *107*–110
strong argument, 6
  vs. valid argument, 7
strong causal inference can be good?, 21, 37, 42–44, 47, 78
strong inference, *3*
subjective claim describes a cause, 52

subjunctive conditional, 112n, 121, 124–125
Sufficient Reason, Principle of, *66*

tautology, *106*
Taylor, Richard, *58, 60, 66*
teleological explanation, *169*–172
  necessary conditions to be good, 170
teleological fallacy, *170*
testable claim, 137. *See also* confirming an explanation; falsifying a prediction
Thagard, Paul R., 153n
theories, 166–169
things as causes, 16, 17, 23, 40, 42, 46, 47–48, 60, 64, 69, 89
  *See also* causal power; nothing
things as objects of emotions, 97–99
Thode, E. Wayne, 28n
thoughts, 1
time and space in causal reasoning, 21–25, 54–55
  cause close to effect in time and space, 25. *See also* tracing cause backwards
time, arrow of, 22, 61
Tooley, Michael *29n*, 55n
Torricelli, 147–148
tracing cause backwards, 32, 37, 50–51, 56

transitivity of cause and effect?, 55
turtle example, 36–37, 38

uncontrolled experiment, 72
Unrepairable Arguments, *9*
Unrepairable Explanations, *133*
untestable explanation, 137
Unwin, Nicholas, 85n
uttering, 1

vagueness, 1–2
valid argument, 6
  vs. strong argument, 7
valid inference, *3*
van der Steen, Wim J., *139*
van Fraassen, Bas C. 129n, 135n, 156n
Vlastos, Gregory, *82*
von Wright, G.H., *60*

Waismann, Friedrich, *44–45n*
Wallace, William A., *40n, 41n*, 77n, 83n
Walters, R.S., 121n, *122*
weak inference, *3*
White, Morton, 78, *92*
Whitehead, Alfred North, 89n
Wright, Larry, 172n

www.ingramcontent.com/pod-product-compliance
Lightning Source LLC
Chambersburg PA
CBHW061305110428
42742CB00012BA/2058